云计算工程师系列

大型网站架构与自动化运维

主 编 肖 睿 罗保山 刘丽军

中国水利水电出版社
www.waterpub.com.cn
·北京·

内 容 提 要

本书针对具备 Linux 基础的人群，主要介绍了缓存代理、高性能内存对象缓存 Memcached、分布式文件服务、大型网站架构、自动化运维的相关知识与应用，以企业级的实战项目案例，使读者能够掌握应用运维的工作内容。项目案例包括 MFS 分布式文件系统、百万/千万 PV 网站架构、Ansible、SaltStack、Puppet 自动化运维，通过以上项目案例的训练，读者能够理解大型网站的架构，达到运维自动化的高度。

本书通过通俗易懂的原理及深入浅出的案例，并配以完善的学习资源和支持服务，为读者带来全方位的学习体验，包括视频教程、案例素材下载、学习交流社区、讨论组等终身学习内容，更多技术支持请访问课工场 www.kgc.cn。

图书在版编目（CIP）数据

大型网站架构与自动化运维 / 肖睿，罗保山，刘丽军主编. -- 北京：中国水利水电出版社，2017.5（2024.9重印）
（云计算工程师系列）
ISBN 978-7-5170-5404-7

Ⅰ. ①大… Ⅱ. ①肖… ②罗… ③刘… Ⅲ. ①网站—建设 Ⅳ. ①TP393.092.1

中国版本图书馆CIP数据核字(2017)第105392号

策划编辑：石永峰　责任编辑：张玉玲　加工编辑：赵佳琦　封面设计：梁　燕

书　　名	云计算工程师系列 大型网站架构与自动化运维 DAXING WANGZHAN JIAGOU YU ZIDONGHUA YUNWEI
作　　者	主编　肖　睿　罗保山　刘丽军
出版发行	中国水利水电出版社 （北京市海淀区玉渊潭南路1号D座 100038） 网　址：www.waterpub.com.cn E-mail：mchannel@263.net（答疑） 　　　　sales@mwr.gov.cn 电　话：（010）68545888（营销中心）、82562819（组稿）
经　　售	北京科水图书销售有限公司 电话：（010）68545874、63202643 全国各地新华书店和相关出版物销售网点
排　　版	北京万水电子信息有限公司
印　　刷	三河市德贤弘印务有限公司
规　　格	184mm×260mm　16开本　11.5印张　266千字
版　　次	2017年5月第1版　2024年9月第3次印刷
印　　数	6001—7000 册
定　　价	36.00 元

凡购买我社图书，如有缺页、倒页、脱页的，本社营销中心负责调换
版权所有·侵权必究

丛书编委会

主　　任：肖　睿

副 主 任：习景涛

委　　员：杨　欢　　潘贞玉　　张德平　　相洪波　　谢伟民
　　　　　庞国广　　张惠军　　段永华　　李　娜　　孙　苹
　　　　　董泰森　　曾谆谆　　王俊鑫　　俞　俊

课工场：李超阳　　祁春鹏　　祁　龙　　滕传雨　　尚永祯
　　　　　张雪妮　　吴宇迪　　曹紫涵　　吉志星　　胡杨柳依
　　　　　李晓川　　黄　斌　　宗　娜　　陈　璇　　王博君
　　　　　习志星　　孙　敏　　张　智　　董文治　　霍荣慧
　　　　　刘景元　　袁娇娇　　李　红　　孙正哲　　史爱鑫
　　　　　周士昆　　傅　峥　　于学杰　　何娅玲　　王宗娟

前　　言

"互联网＋人工智能"时代，新技术的发展可谓是一日千里，云计算、大数据、物联网、区块链、虚拟现实、机器学习、深度学习等等，已经形成一波新的科技浪潮。以云计算为例，国内云计算市场的蛋糕正变得越来越诱人，以下列举了2016年以来发生的部分大事。

1. 中国联通发布云计算策略，并同步发起成立"中国联通沃云＋云生态联盟"，全面开启云服务新时代。

2. 内蒙古斥资500亿元欲打造亚洲最大云计算数据中心。

3. 腾讯云升级为平台级战略，旨在探索云上生态，实现全面开放，构建可信赖的云生态体系。

4. 百度正式发布"云计算＋大数据＋人工智能"三位一体的云战略。

5. 亚马逊AWS和北京光环新网科技股份有限公司联合宣布：由光环新网负责运营的AWS中国（北京）区域在中国正式商用。

6. 来自Forrester的报告认为，AWS和OpenStack是公有云和私有云事实上的标准。

7. 网易正式推出"网易云"。网易将先行投入数十亿人民币，发力云计算领域。

8. 金山云重磅发布"大米"云主机，这是一款专为创业者而生的性能王云主机，采用自建11线BGP全覆盖以及VPC私有网络，全方位保障数据安全。

DT时代，企业对传统IT架构的需求减弱，不少传统IT企业的技术人员，面临失业风险。全球最知名的职业社交平台LinkedIn发布报告，最受雇主青睐的十大职业技能中"云计算"名列前茅。2016年，中国企业云服务整体市场规模超500亿元，预计未来几年仍将保持约30%的年复合增长率。未来5年，整个社会对云计算人才的需求缺口将高达130万。从传统的IT工程师转型为云计算与大数据专家，已经成为一种趋势。

基于云计算这样的大环境，课工场（www.kgc.cn）的教研团队几年前开始策划的"云计算工程师系列"教材应运而生，它旨在帮助读者朋友快速成长为符合企业需求的、优秀的云计算工程师。这套教材是目前业界最全面、专业的云计算课程体系，能够满足企业对高级复合型人才的要求。参与编写的院校老师还有罗保山、刘丽军等。

课工场是北京大学下属企业北京课工场教育科技有限公司推出的互联网教育平台，专注于互联网企业各岗位人才的培养。平台汇聚了数百位来自知名培训机构、高校的顶级名师和互联网企业的行业专家，面向大学生以及需要"充电"的在职人员，针对与互联网相关的产品设计、开发、运维、推广和运营等岗位，提供在线的直播和录播课程，并通过遍及全国的几十家线下服务中心提供现场面授以及多种形式的教学服务，并同步研发出版最新的课程教材。

除了教材之外，课工场还提供各种学习资源和支持服务，包括：

- 现场面授课程
- 在线直播课程
- 录播视频课程
- 授课PPT课件
- 案例素材下载
- 扩展资料提供
- 学习交流社区
- QQ讨论组（技术，就业，生活）

以上资源请访问课工场网站 www.kgc.cn。

本套教材特点

（1）科学的训练模式
- 科学的课程体系。
- 创新的教学模式。
- 技能人脉，实现多方位就业。
- 随需而变，支持终身学习。

（2）企业实战项目驱动
- 覆盖企业各项业务所需的IT技能。
- 几十个实训项目，快速积累一线实践经验。

（3）便捷的学习体验
- 提供二维码扫描，可以观看相关视频讲解和扩展资料等知识服务。
- 课工场开辟教材配套版块，提供素材下载、学习社区等丰富的在线学习资源。

读者对象

（1）初学者：本套教材将帮助你快速进入云计算及运维开发行业，从零开始逐步成长为专业的云计算及运维开发工程师。

（2）初中级运维及运维开发者：本套教材将带你进行全面、系统的云计算及运维开发学习，逐步成长为高级云计算及运维开发工程师。

课工场出品（www.kgc.cn）

课程设计说明

课程目标

读者学完本书后,能够掌握自动化运维技术,设计、实施和维护大型网站架构。

训练技能

- 理解缓存代理的工作原理,并掌握其配置。
- 理解高性能内存对象缓存 Memcached 的工作原理,并掌握其部署。
- 掌握异地备份工具 rsync 的工作原理及其相关配置。
- 掌握 MFS 工作原理、部署与灾难恢复。
- 理解百万、千万 PV 网站架构及其相关部署。
- 理解自动化运维工具原理并进行自动化配置。

设计思路

本书采用了教材+扩展知识的设计思路,扩展知识提供二维码扫描,形式可以是文档、视频等,内容可以随时更新,能够更好地服务读者。

教材分为 3 个阶段来设计学习过程,即代理与缓存、网站与容灾、自动化运维,具体安排如下:

- 第 1 章~第 2 章介绍缓存代理与内存缓存的相关知识,理解 Squid 代理的工作模式、Memcached 的工作原理,实现 Squid 代理和 Memcached 相关部署。
- 第 3 章~第 6 章是构建企业网站相关内容,使用 rsync 技术实现企业网站的备份,MFS 实现灾难恢复与在线扩容,进而实现百万、千万 PV 网站架构部署与管理。
- 第 7 章~第 9 章介绍企业自动化运维工具 Ansible、SaltStack、Puppet 的部署与使用。

章节导读

- 技能目标:学习本章所要达到的技能,可以作为检验学习效果的标准。
- 本章导读:对本章涉及的技能内容进行分析并展开讲解。
- 操作案例:对所学内容的实操训练。
- 本章总结:针对本章内容的概括和总结。
- 本章作业:针对本章内容的补充练习,用于加强对技能的理解和运用。

- 扩展知识：针对本章内容的扩展、补充，对于新知识随时可以更新。

学习资源

- 学习交流社区（课工场）
- 案例素材下载
- 相关视频教程

更多内容详见课工场 www.kgc.cn。

目　　录

前言
课程设计说明

第 1 章　Squid 缓存服务器 1
1.1　Squid 服务基础 2
1.1.1　缓存代理概述 2
1.1.2　安装及运行控制 3
1.2　构建代理服务器 6
1.2.1　传统代理 6
1.2.2　透明代理 9
1.2.3　ACL 访问控制 11
1.3　Squid 日志分析 13
1.4　Squid 反向代理 15
1.5　Varnish 与 Nginx 缓存服务器 16
本章总结 17
本章作业 17

第 2 章　高性能内存对象缓存 Memcached 19
2.1　认识 Memcached 20
2.2　安装 Memcached 案例 23
2.2.1　安装 Memcached 服务器 ... 23
2.2.2　Memcached API 客户端 24
2.3　Memcached 数据库操作与管理 27
2.4　Memcached 实现主主复制和高可用的方式 30
2.4.1　Memcached 主主复制架构 ... 30
2.4.2　Memcached 主主复制 +Keepalived 高可用架构 32
本章总结 35

第 3 章　rsync 远程同步 37
3.1　配置 rsync 源服务器 38
3.2　使用 rsync 备份工具 40
3.3　配置 rsync+inotify 实时同步ceil 41
本章总结 44
本章作业 44

第 4 章　MFS 分布式文件系统 45
4.1　案例分析 46
4.2　案例实施 48
本章总结 59
本章作业 59

第 5 章　部署社交网站 61
5.1　案例分析 62
5.1.1　案例概述 62
5.1.2　案例环境 62
5.2　案例实施 63
5.3　案例扩展 67
本章总结 67

第 6 章　大型网站架构 69
6.1　网站架构概述 70
6.2　百万 PV 网站架构案例 70
6.3　千万 PV 网站架构案例 81
6.3.1　案例概述 81
6.3.2　RabbitMQ 集群配置 82
本章总结 95

第 7 章　自动化运维之 Ansible 97
7.1　Ansible 概述 98
7.2　Ansible 核心组件 99
7.3　安装部署 Ansible 服务 100
7.4　Ansible 命令应用基础 102
7.5　YAML 介绍 114
7.6　Ansible 基础元素介绍 115
7.7　Playbook 介绍 119
本章总结 136

第 8 章　自动化运维之 SaltStack .. 137
8.1　SaltStack 基础 138
8.2　SaltStack 批量部署并配置 Nginx 139
8.2.1　需求分析 139
8.2.2　操作步骤 139
本章总结 149

第 9 章　自动化运维之 Puppet 151
9.1　案例一分析 152
9.2　案例一实施 154
9.3　案例二分析 162
9.4　案例二实施 163
本章总结 173
本章作业 174

第 1 章

Squid 缓存服务器

技能目标

- 学会构建传统代理、透明代理、反向代理服务
- 学会配置 Squid 的访问控制策略
- 熟悉 Varnish、Nginx 缓存服务器

本章导读

Squid 是 Linux 系统中最常用的一款开源代理服务软件（官方网站为 http://www.squid-cache.org），可以很好地实现 HTTP、FTP、DNS 查询，以及 SSL 等应用的缓存代理，功能十分强大。

知识服务

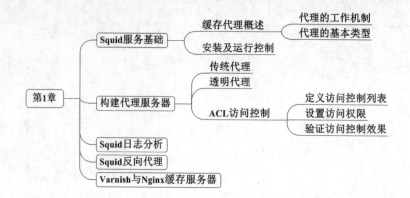

1.1 Squid 服务基础

本节将介绍缓存代理的工作机制、类型,以及 Squid 服务的安装、运行控制和配置文件。

1.1.1 缓存代理概述

作为应用层的代理服务软件,Squid 主要提供缓存加速和应用层过滤控制的功能。

1. 代理的工作机制

当客户机通过代理来请求 Web 页面时,指定的代理服务器会先检查自己的缓存,如果缓存中已经有客户机需要访问的页面,则直接将缓存中的页面内容反馈给客户机;如果缓存中没有客户机需要访问的页面,则由代理服务器向 Internet 发送访问请求,当获得返回的 Web 页面以后,将网页数据保存到缓存中并发送给客户机,如图 1.1 所示。

图 1.1 代理服务的缓存机制

HTTP 代理的缓存加速对象主要是文字、图像等静态 Web 元素。使用缓存机制后,当客户机在不同的时候访问同一 Web 元素,或者不同的客户机访问相同的 Web 元素时,可以直接从代理服务器的缓存中获得结果。这样就大大减少了向 Internet 重复提交 Web 请求的过程,提高了客户机的 Web 访问响应速度。

由于客户机的 Web 访问请求实际上是由代理服务器来代替完成的,从而可以隐藏用户的真实 IP 地址，起到一定的保护作用。另一方面，代理服务器担任着类似"经纪人"的角色，所以有机会针对要访问的目标、客户机的地址、访问的时间段等进行过滤控制。

2. 代理的基本类型

根据实现的方式不同，代理服务可分为传统代理和透明代理两种常见的代理服务。

- 传统代理：也就是普通的代理服务，首先必须在客户机的浏览器、QQ 聊天工具、下载软件等程序中手动设置代理服务器的地址和端口，然后才能使用代理服务来访问网络。对于网页浏览器，访问网站时的域名解析请求也会发送给指定的代理服务器。
- 透明代理：提供与传统代理相同的功能和服务，其区别在于客户机不需要指定代理服务器的地址和端口，而是通过默认路由、防火墙策略将 Web 访问重定向，实际上仍然交给代理服务器来处理。重定向的过程对客户机来说是"透明"的，用户甚至并不知道自己在使用代理服务，所以称为"透明代理"。使用透明代理时，网页浏览器访问网站时的域名解析请求将优先发给 DNS 服务器。

在实际应用中，传统代理多见于 Internet 环境，如为 QQ 程序使用代理可以隐藏本机真实 IP 地址，为下载工具使用多个代理可以规避服务器的并发连接限制。而透明代理多见于局域网环境，如在 Linux 网关中启用透明代理后，局域网主机无需进行额外的设置就可以享受更好的上网速度。

1.1.2 安装及运行控制

下面以 Squid 3.4.6 版为例，介绍其安装和运行控制。

1. 编译安装 Squid

配置 Squid 的编译选项时，将安装目录设置为 /usr/local/squid，其他具体选项根据实际需要来确定，配置前可参考 ./configure --help 给出的说明。

```
[root@localhost ~]# tar zxf squid-3.4.6.tar.gz
[root@localhost ~]# cd squid-3.4.6
[root@localhost squid-3.4.6]# ./configure --prefix=/usr/local/squid
    -- sysconfdir =/etc --enable-arp-acl --enable-linux-netfilter --enable-linux-tproxy
    --enable-async-io=100 --enable-err-language="Simplify_Chinese"
    --enable-underscore --enable-poll --enable-gnuregex
[root@localhost squid-3.4.6]# make && make install
```

上述选项含义如下：

```
--prefix=/usr/local/squid                      // 安装目录
--sysconfdir=/etc                              // 单独将配置文件修改到其他目录
--enable-arp-acl
// 可以在规则中设置为直接通过客户端 MAC 进行管理，防止客户端使用 IP 欺骗
--enable-linux-netfilter                       // 使用内核过滤
--enable-linux-tproxy                          // 支持透明模式
```

```
--enable-async-io= 值
// 异步 I/O，提升存储性能，相当于 --enable-pthreads  -- enable-storeio=ufs,aufs
// --with -pthreads --with-aufs-thread= 值
--enable-err-language="Simplify_Chinese"      // 错误信息的显示语言
--enable-underscore                           // 允许 URL 中有下划线
--enable-poll                                 // 使用 Poll() 模式，提升性能
--enable-gnuregex                             // 使用 GNU 正则表达式
```

安装完后，创建链接文件、创建用户和组。

```
[root@localhost ~]# ln -s /usr/local/squid/sbin/* /usr/local/sbin/
[root@localhost ~]# useradd -M -s /sbin/nologin squid
[root@localhost ~]# chown -R squid:squid /usr/local/squid/var/
```

2. Squid 的配置文件

Squid 服务的配置文件位于 /etc/squid.conf，充分了解配置行的作用将有助于管理员根据实际情况灵活地配置代理服务。更详细的配置项请参考 /etc/squid.conf.documented 文件。

```
acl localnet src 10.0.0.0/8          # RFC1918 possible internal network
acl localnet src 172.16.0.0/12       # RFC1918 possible internal network
acl localnet src 192.168.0.0/16      # RFC1918 possible internal network
acl localnet src fc00::/7            # RFC 4193 local private network range
acl localnet src fe80::/10           # RFC 4291 link-local (directly plugged) machines
acl SSL_ports port 443
acl Safe_ports port 80               # http
acl Safe_ports port 21               # ftp
acl Safe_ports port 443              # https
acl Safe_ports port 70               # gopher
acl Safe_ports port 210              # wais
acl Safe_ports port 1025-65535       # unregistered ports
acl Safe_ports port 280              # http-mgmt
acl Safe_ports port 488              # gss-http
acl Safe_ports port 591              # filemaker
acl Safe_ports port 777              # multiling http
acl CONNECT method CONNECT
http_access deny !Safe_ports
http_access deny CONNECT !SSL_ports
http_access allow localhost manager
http_access deny manager
http_access allow localnet
http_access allow localhost
http_access deny all
http_port 3128                       // 用来指定代理服务监听的地址和端口（默认的端口号为 3128）
cache_effective_user squid           // 这一项指定 squid 的程序用户，用来设置初始化、运行时缓存
                                     // 的账号，否则启动不成功！
cache_effective_group squid          // 默认为 cache_effective_user 指定账号的基本组
coredump_dir /usr/local/squid/var/cache/squid
```

refresh_pattern ^ftp:	1440	20%	10080
refresh_pattern ^gopher:	1440	0%	1440
refresh_pattern -i (/cgi-bin/|\?) 0	0%	0	
refresh_pattern .	0	20%	4320

3. Squid 的运行控制

（1）检查配置文件的语法是否正确。

[root@localhost squid]# **squid –k parse**

（2）启动、停止 Squid。

第一次启动 Squid 服务时，会自动初始化缓存目录。在没有可用的 Squid 服务脚本的情况下，也可以直接调用 Squid 程序来启动服务，这时需要先进行初始化。

[root@localhost ~]# **squid -z**　　　　　//-z 选项用来初始化缓存目录
[root@localhost ~]# **squid**　　　　　　// 启动 squid 服务

确认 Squid 服务处于正常监听状态。

[root@localhost ~]# netstat -anpt | grep "squid"
tcp   0   0 0.0.0.0:3128   0.0.0.0:*   LISTEN   9947/(squid).

（3）使用 Squid 服务脚本。

为了使 Squid 服务的启动、停止、重载等操作更加方便，可以编写 Squid 服务脚本，并使用 chkconfig 和 service 工具来进行管理。

```bash
#!/bin/bash
# chkconfig: 2345 90 25
# config: /etc/squid.conf
# pidfile: /usr/local/squid/var/run/squid.pid
# Description: Squid - Internet Object Cache.
PID="/usr/local/squid/var/run/squid.pid"
CONF="/etc/squid.conf"
CMD="/usr/local/squid/sbin/squid"
case "$1" in
   start)
      netstat -anpt | grep squid   &> /dev/null
      if [ $? -eq 0 ]
      then
         echo "squid is running"
       else
         echo " 正在启动 squid..."
         $CMD
      fi
   ;;
   stop)
      $CMD -k kill &> /dev/null
      rm -rf $PID  &> /dev/null
   ;;
```

```
        status)
            [ -f $PID ] &> /dev/null
            if [ $? -eq 0 ]
               then
            netstat -anpt | grep squid
               else
                  echo "Squid is not running."
               fi
        ;;
        restart)
            $0 stop  &> /dev/null
            echo " 正在关闭 squid..."
               $0 start  &> /dev/null
            echo " 正在启动 squid..."
        ;;
        reload)
            $CMD -k reconfigure
        ;;
        check)
            $CMD -k parse
        ;;
        *)
            echo " 用法： $0 {start | stop | restart | reload | check | status}"
        ;;
        esac
        [root@localhost ~]# chmod +x /etc/init.d/squid
        [root@localhost ~]# chkconfig --add squid          // 添加为系统服务
        [root@localhost ~]# chkconfig squid on
```

这样一来，就可以通过 Squid 脚本来启动、停止、重启和重载 Squid 服务器了，方法是在执行时添加相应的 start、stop、restart 和 reload 参数。

1.2 构建代理服务器

本节主要从三个方面来学习 Squid 服务的构建和使用，分别为传统代理、透明代理和访问控制列表（ACL）。传统代理的实现最为简单，透明代理还需要结合默认路由、防火墙策略等一起来完成，访问控制列表主要用来针对客户机的 Web 访问过程进行过滤控制。

1.2.1 传统代理

使用传统代理的特点在于，客户机的相关程序（如 IE 浏览器、QQ 聊天软件等）必须指定代理服务器的地址、端口等基本信息。下面通过一个简单的应用案例来学习传统代理的配置和使用。

基于 Internet 网络环境，如图 1.2 所示，案例的主要需求描述如下。
- 在 Linux 主机 B 上，构建 Squid 为客户机访问各种网站提供代理服务，但禁止通过代理下载超过 10MB 大小的文件。
- 在客户机 C 上，指定主机 B 作为 Web 访问代理，以隐藏自己的真实 IP 地址。

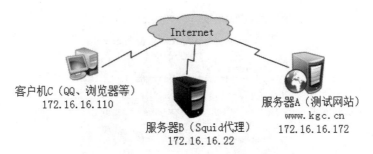

图 1.2 使用 Squid 传统代理

针对上述实验环境，主机 B 作为代理服务器，必须正确构建 Squid 服务，并允许客户机使用代理。若要客户机通过代理以 http://www.kgc.cn/ 的域名形式进行访问，则代理服务器本身必须能够正确解析该域名。主机 C 作为客户机，需要为浏览器等程序指定所使用的代理服务器地址、端口号等信息。主机 A 作为测试网站，需要启用 httpd 服务。

下面主要介绍 Squid 服务器的配置、客户机的代理设置，以及代理服务的验证方法。

1．Squid 服务器的配置

配置 Squid 实现传统代理服务时，需要注意添加 http_access allow all 访问策略，以便允许任意客户机使用代理服务。除此以外，为了限制下载文件的大小，还需要设置 reply_body_max_size 项，其他各种参数均可保持默认。

（1）修改 squid.conf 配置文件。

```
[root@localhost ~]# vi /etc/squid.conf
http_port 3128
reply_body_max_size 10 MB          // 允许下载的最大文件大小（10MB）
http_access allow all              // 放在 http_access deny all 之前
……                                  // 省略部分内容
```

在防火墙上添加允许策略：

```
[root@localhost ~]# iptables -I INPUT -p tcp --dport 3128 -j ACCEPT
[root@localhost ~]# service iptables save
iptables：将防火墙规则保存到 /etc/sysconfig/iptables：      [ 确定 ]
```

（2）重载 Squid 服务。

修改 squid.conf 配置文件以后，需要重新加载方可生效。执行 service squid reload 或者 squid -k reconfigure，都可以重新加载服务配置。

```
[root@localhost ~]# service squid reload
```

2. 客户机的代理配置

在 IE 浏览器中，选择"工具"→"Internet 选项"，弹出"Internet 选项"对话框，在"连接"选项卡中的"局域网（LAN）设置"选项组中单击"局域网设置"按钮，弹出"局域网（LAN）设置"对话框，如图 1.3 所示。

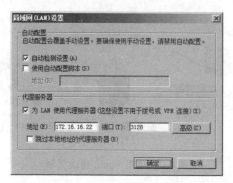

图 1.3 IE 浏览器的代理设置

若要在 Linux 客户机的命令行界面中使用代理服务器（如 elinks 网页浏览器、wget 下载工具），必须通过环境变量来指定代理服务器的地址、端口等信息。

```
[root@localhost ~]# vi /etc/profile
……                                              // 省略部分内容
HTTP_PROXY=http://172.16.16.22:3128             // 为使用 HTTP 协议指定代理
HTTPS_PROXY=http://172.16.16.22:3128            // 为使用 HTTPS 协议指定代理
FTP_PROXY=http://172.16.16.22:3128              // 为使用 FTP 协议指定代理
NO_PROXY=192.168.1.,192.168.4.                  // 对两个局域网段不使用代理
export HTTP_PROXY HTTPS_PROXY FTP_PROXY NO_PROXY
[root@localhost ~]# source /etc/profile
```

3. 代理服务的验证方法

在客户机 172.16.16.110 中通过浏览器访问目标网站 http://172.16.16.172/，然后观察 Squid 代理服务器、Web 服务器的访问日志，以验证代理服务是否发挥作用。

（1）查看 Squid 访问日志的新增记录。

在 Squid 代理服务器中，通过跟踪 Squid 服务的访问日志文件，应该能够发现客户机 172.16.16.110 访问网站服务器 172.16.16.172 的记录。

```
[root@localhost ~]# tail /usr/local/squid/var/logs/access.log
……                                              // 省略部分内容
1309238261.011    34 172.16.16.110 TCP_MISS/200 459 GET http://172.16.16.172/
    - DIRECT/172.16.16.172 text/html
1309238261.126   113 172.16.16.110 TCP_MISS/404 628 GET http://172.16.16.
    172/favicon.ico - DIRECT/172.16.16.172 text/html
```

（2）查看 Web 访问日志的新增记录。

在被访问的 Web 服务器中，通过跟踪 httpd 服务的访问日志文件，应该能够发现

来自代理服务器 172.16.16.22 的访问记录。这说明当客户机使用代理以后，Web 服务器并不知道客户机的真实 IP 地址，因为实际上是由代理服务器在替它访问。

```
[root@localhost ~]# tail /var/log/httpd/access_log
172.16.16.22 - - [18/Jul/2016:08:46:59 +0800] "GET / HTTP/1.1" 403 5039 "-"
"Mozilla/4.0 (compatible; MSIE 8.0; Windows NT 6.1; WOW64; Trident/4.0; SLCC2; .NET
CLR 2.0.50727; .NET CLR 3.5.30729; .NET CLR 3.0.30729; Media Center PC 6.0)"
```

当客户机再次访问同一 Web 页面时，Squid 访问日志中会增加新的记录，但 Web 访问日志中的记录不会有变化（除非页面变更或执行强制刷新等操作）。这说明当客户机重复访问同一静态页面时，实际上页面是由代理服务器通过缓存提供的。

1.2.2 透明代理

透明代理提供的服务功能与传统代理是一致的，但是其"透明"的实现依赖于默认路由和防火墙的重定向策略，因此更适用于为局域网主机服务，而不适合为 Internet 中的客户机提供服务。下面也通过一个简单的应用案例来学习透明代理的配置和使用。

基于局域网主机通过 Linux 网关访问 Internet 的环境，如图 1.4 所示，案例的主要需求描述如下。

- 在 Linux 网关上，构建 Squid 为客户机访问 Internet 提供代理服务。
- 在所有的局域网客户机上，只需正确设置 IP 地址、默认网关，而不需要手动指定代理服务器的地址、端口等信息。

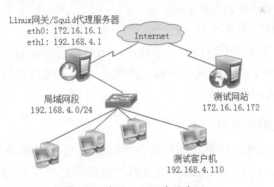

图 1.4　使用 Squid 透明代理

针对上述实验环境，透明代理的关键在于 Linux 网关服务器，而对于客户机仅需正确地设置网络地址、默认网关，并不需要指定代理服务器（若指定了反而易出错）。

关于客户机的 DNS 解析工作，最好还是通过正常的 DNS 服务器来提供，不建议抛给代理服务器来处理。下面主要介绍 Squid 服务的透明代理设置、防火墙策略设置，其他配置操作请参考前面的传统代理构建过程。

1. 配置 Squid 支持透明代理

Squid 服务的默认配置并不支持透明代理，因此需要调整相关设置。对于 2.6 以上版本的 Squid 服务，只要在 http_port 配置行加上一个 transparent（透明）选项，就可以支持透明代理了。

```
[root@localhost ~]# vi /etc/squid.conf
……                                    // 省略部分内容
http_port 192.168.4.1:3128 transparent // 只在其中一个 IP 地址上提供服务
[root@localhost ~]# service squid reload
```

2. 设置 iptables 的重定向策略

透明代理中的 Squid 服务实际上是构建在 Linux 网关主机上的，因此只需正确设置防火墙策略，就可以将局域网主机访问 Internet 的数据包转交给 Squid 进行处理。这需要用到 iptables 的 REDIRECT（重定向）策略，其作用是实现本机端口的重新定向，将访问网站协议 HTTP、HTTPS 的外发数据包转交给本机的 Squid 服务（3128 端口）。

REDIRECT 也是一种数据包控制类型，只能在 nat 表的 PREROUTING 或 OUTPUT 链以及被其调用的链中使用，通过 "--to-ports 端口号"的形式来指定映射的目标端口。本例中可以将来自局域网段 192.168.4.0/24 且访问 HTTP、HTTPS 等协议的数据包转交给运行在本机 3128 端口上的 Squid 服务进行处理。

防火墙做重定向操作，将访问本机 80、443 端口的请求重定向到 3128 端口。

```
[root@localhost ~]# iptables -t nat -I PREROUTING -i eth1 -s 192.168.10.0/24
    -p tcp --dport 80 -j REDIRECT --to 3128
[root@localhost ~]# iptables -t nat -I PREROUTING -i eth1 -s 192.168.10.0/24
    -p tcp --dport 443 -j REDIRECT --to 3128
[root@localhost ~]# service iptables save
iptables: 将防火墙规则保存到 /etc/sysconfig/iptables：    [ 确定 ]
```

由于 FTP 协议涉及多个端口、多个连接，虽然也可以通过 HTTP 代理进行访问，但使用透明代理不便实现，因此最佳做法仍然是采用传统代理的方式——手动指定代理服务器的地址、端口号。

3. 验证透明代理的使用

为了验证透明代理的效果，如果存在手动指定的代理服务器设置，应在客户机中将其去除。例如，在 IE 或 Firefox 浏览器的连接设置中不要勾选使用代理服务器；在 Linux 客户机的命令行界面中，可以通过 unset 命令清除 HTTP_PROXY、HTTPS_PROXY 等变量。

```
[root@localhost ~]# unset HTTP_PROXY HTTPS_PROXY
```

在客户机 192.168.4.110 中通过浏览器访问目标网站 http://172.16.16.172/，然后观察 Squid 代理服务器、Web 服务器的访问日志，以验证透明代理是否发挥作用。操作方法可参考 1.2.1 节，验证结果为在 Squid 代理服务器中，应该能够发现客户机 192.168.4.110 访问网站服务器 172.16.16.172 的记录；在被访问的 Web 服务器中，应该能够发现来自代理服务器 172.16.16.1 的访问记录。

1.2.3 ACL 访问控制

Squid 提供了强大的代理控制机制，通过合理设置 ACL（Access Control List，访问控制列表）并进行限制，可以针对源地址、目标地址、访问的 URL 路径、访问的时间等各种条件进行过滤。

在配置文件 squid.conf 中，ACL 访问控制通过以下两个步骤来实现：其一，使用 acl 配置项定义需要控制的条件；其二，通过 http_access 配置项对已定义的列表做"允许"或"拒绝"访问的控制。

1．定义访问控制列表

每一行 acl 配置可以定义一条访问控制列表，格式如下：

acl 列表名称 列表类型 列表内容 …

其中，"列表名称"由管理员自行指定，用来识别控制条件；"列表类型"必须使用 Squid 预定义的值，对应不同类别的控制条件；"列表内容"是要控制的具体对象，不同类型的列表所对应的内容也不一样，可以有多个值（以空格分隔，都为"或"的关系）。

通过上述格式可以发现，定义访问控制列表时，关键在于选择"列表类型"并设置具体的条件对象。Squid 预定义的列表类型有很多种，常用的包括源地址、目标地址、访问时间、访问端口等，如表 1-1 所示。

表 1-1 常用的访问控制列表类型

列表类型	列表内容示例	含义 / 用途
src	192.168.1.168 192.168.1.0/24 192.168.1.0-192.168.3.0/24	源 IP 地址、网段、IP 地址范围
dst	216.163.137.3 61.135.167.0/24 www.playboy.com	目标 IP 地址、网段、主机名
port	80 443 8080 20 21	目标端口
dstdomain	.qq.com	目标域，匹配域内的所有站点
time	MTWHF 8:30-17:30 12:00-13:00 AS	使用代理服务的时间段 字母表示一星期中各天的英文缩写 M—Monday、T—Tuesday、W—Wednesday、H—Thursday、F—Friday、A—Saturday、S—Sunday
maxconn	20	每个客户机的并发连接数
url_regex	url_regex -i ^rtsp:// url_regex -i ^emule://	目标资源的 URL 地址，-i 表示忽略大小写
urlpath_regex	urlpath_regex -i sex adult urlpath_regex -i \.mp3$	目标资源的整个 URL 路径，-i 表示忽略大小写

在定义访问控制列表时，应结合当前网络环境正确分析用户的访问需求，准确定义使用代理服务的控制条件。例如，针对不同的客户机地址、需要限制访问的目标网站、特定的时间段……，分别定义列表。

```
[root@localhost ~]# vi /etc/squid.conf
……                                                       // 省略部分内容
acl localhost src 127.0.0.1/255.255.255.255               // 源地址为 127.0.0.1
acl MYLAN src 192.168.1.0/24 192.168.4.0/24               // 客户机网段
acl to_localhost dst 127.0.0.0/8                          // 目标地址为 127.0.0.0/8 网段
acl MC20 maxconn 20                                       // 最大并发连接 20
acl BlackURL url_regex -i ^rtsp:// ^emule://              // 以 rtsp:// 等开头的 URL
acl MEDIAFILE urlpath_regex -i \.mp3$ \.mp4$ \.rmvb$
                                                          // 以 .mp3、.mp4、.rmvb 结尾的 URL 路径
acl WORKTIME time MTWHF 08:30-17:30                       // 时间为周一至周五 8:30~17:30
```

当需要限制的同一类对象较多时，可以使用独立的文件来存放，在 acl 配置行的列表内容处指定对应的文件位置即可。例如，若要针对目标地址建立黑名单文件，可以参考以下操作。

```
[root@localhost ~]# mkdir /etc/squid
[root@localhost ~]# cd /etc/squid
[root@localhost squid]# vi ipblock.list                   // 建立目标 IP 地址名单
61.135.167.36
125.39.127.25
60.28.14.0/24
[root@localhost squid]# vi dmblock.list                   // 建立目标域地址名单
.qq.com
.msn.com
.live.com
[root@localhost squid]# vi /etc/squid.conf
……                                                       // 省略部分内容
acl IPBLOCK dst "/etc/squid/ipblock.list"                 // 调用指定文件中的列表内容
acl DMBLOCK dstdomain "/etc/squid/dmblock.list"
```

2. 设置访问权限

定义好各种访问控制列表以后，需要使用 httpd_access 配置项来进行控制。需要注意的是，http_access 配置行必须放在对应的 acl 配置行之后。每一行 http_access 配置确定一条访问控制规则，格式如下：

```
http_access allow 或 deny 列表名……
```

在每一条 http_access 规则中，可以同时包含多个访问控制列表名，各个列表之间以空格分隔，为"与"的关系，表示必须满足所有访问控制列表对应的条件才会进行限制。需要使用取反条件时，可以在访问控制列表前添加"!"符号。

```
[root@localhost ~]# vi /etc/squid.conf
……                                                       // 省略部分内容
```

```
http_access deny MYLAN MEDIAFILE          // 禁止客户机下载 MP3、MP4 等文件
http_access deny MYLAN IPBLOCK            // 禁止客户机访问黑名单中的 IP 地址
http_access deny MYLAN DMBLOCK            // 禁止客户机访问黑名单中的网站域
http_access deny MYLAN MC20               // 客户机的并发连接超过 20 时将被阻止
http_access allow MYLAN WORKTIME          // 允许客户机在工作时间上网
http_access deny all                      // 默认禁止所有客户机使用代理
```

执行访问控制时，Squid 将按照各条规则的顺序依次进行检查，如果找到一条相匹配的规则就不再向后搜索（这点与 iptables 的规则匹配类似）。因此，规则的顺序安排是非常重要的，以下两种默认情况需要我们注意。

- 没有设置任何规则时：Squid 服务将拒绝客户端的请求。
- 有规则但找不到相匹配的项：Squid 将采用与最后一条规则相反的权限，即如果最后一条规则是 allow，就拒绝客户端的请求，否则允许该请求。

通常情况下，把最常用到的控制规则放在最前面，以减少 Squid 的负载。在访问控制的总体策略上，建议采用"先拒绝后允许"或"先允许后拒绝"的方式，最后一条规则设为默认策略，可以为 http_access allow all 或者 http_access deny all。

3. 验证访问控制效果

关于 Squid 服务的访问控制效果，无外乎两种情况：一种是能够正常访问，另一种是禁止访问。当客户机的代理访问请求被 Squid 服务拒绝时，在浏览器中会看到 ERROR 报错页面，具体内容会根据限制条件的不同而有些细小差别。

（1）测试访问权限限制

对于使用 http_access 规则拒绝访问的情况（如访问被禁止的网站或者在禁止的时间段访问），浏览器的报错页面中会出现 Access Denied 的提示。

（2）测试文件下载限制

对于 1.2.1 节限制文件下载大小的情况（reply_body_max_size 配置项），当下载超过指定大小的 Web 对象时，浏览器的报错页面中会出现 The request or reply is too large. 的提示。

用来下载测试的文件可以通过 dd 命令生成。例如，若限制的大小为 10MB，则可以在目标网站服务器中创建一个 15MB 的测试文件。

```
[root@localhost ~]# dd if=/dev/zero of=/var/www/html/dltest.data bs=1M count=15
```

1.3　Squid 日志分析

Sarg 的全名是 Squid Analysis Report Generator，是一款 Squid 日志分析工具，采用 HTML 格式，详细列出每一位用户访问 Internet 的站点信息、时间占用信息、排名、连接次数和访问量等。

Sarg 部署过程如下所述。

1. 安装 GD 库

[root@localhost ~]# yum -y install gd gd-devel

2. 安装 sarg

[root@localhost ~]# mkdir /usr/local/sarg
[root@localhost ~]# tar zxf sarg-2.3.7.tar.gz
[root@localhost ~]# cd sarg-2.3.7
[root@localhost sarg-2.3.7#./configure --prefix=/usr/local/sarg –sysconfdir =/etc/sarg --enable-extraprotection && make && make install

上述配置项含义如下：

--sysconfdir=/etc/sarg	//配置文件目录，默认是 /usr/loca/etc
--enable-extraprotection	//添加额外的安全保护

3. 配置

```
[root@localhost ~]# cd /etc/sarg/
[root@localhost sarg]# vim sarg.conf
access_log /usr/local/squid/var/logs/access.log   // 指定 squid 的访问日志文件
title "Squid User Access Reports"                 // 网页标题
output_dir /var/www/html/sarg                     //sarg 报告的输出目录
user_ip no                                        // 使用用户名显示
exclude_hosts /usr/local/sarg/noreport            // 指定不计入排序的站点列表文件
top_user_sort_field connect BYTES reverse         // 在 top 排序中，指定连接次数、访问字节数，
                                                  // 采用降序排列，升序将 reverse 换成 normal
user_sort_field connect reverse                   // 对于用户访问记录，连接次数按降序排列
overwrite_report no                               // 当那个日期报告已经存在，是否覆盖报告
mail_utility mailq.postfix                        // 发送邮件报告的命令
charset UTF-8                                     // 使用字符集
weekdays 0-6                                      // 指定 top 排序时的星期周期，0 为周日
hours 9-12,14-16,18-20                            // 指定 top 排序时的时间周期
www_document_root /var/www/html                   // 网页根目录
```

4. 运行

上述配置中，添加了不计入排序的站点，需要存在下面这个文件。

[root@localhost sarg]# touch /usr/local/sarg/noreport

在这个文件中添加的域名将不被显示在排序中。

直接执行 sarg 即可启动一次记录，建议设置符号链接，然后执行 sarg，会看到提示信息：

[root@localhost sarg]# ln -s /usr/local/sarg/bin/sarg /usr/local/bin/
[root@localhost sarg]# sarg
SARG: 记录在文件 : 68, reading: 100.00%
SARG: 成功的生成报告在 /var/www/html/sarg/2014Aug06-2014Aug11

5. 验证

在 sarg.conf 配置中可以设置字体、颜色等信息，方便查询。多次执行 sarg 后，在客户端上访问 sarg/ 目录，可以看到生成的报表。

6. 计划任务

可将 sarg 做成计划任务，定期执行。

如下示例，写个每日报告：

```
[root@localhost ~]# vim /usr/local/sarg/daily.sh    // 示例：每日报告，内容如下：
#/bin/bash
#Get current date
TODAY=$(date +%d/%M/%Y)
#Get one week ago today
YESTERDAY=$(date -d "1 day ago" +%d/%m/%Y)
/usr/local/sarg/bin/sarg -l /usr/local/squid/var/logs/access.log -o /var/www/
      html/sarg -z -d $YESTERDAY-$TODAY  &> /dev/null
exit 0

[root@localhost ~]# chmod +x /usr/local/sarg/daily.sh
[root@localhost ~]# crontab -e                      // 添加任务计划，每天 00:00 执行
00 00 * * * /usr/local/sarg/daily.sh
[root@localhost ~]# chkconfig crond on
```

1.4 Squid 反向代理

许多大型的门户网站架构中都采用了反向代理加速，使用比较多的是 nginx、squid 等。

如图 1.5 所示，通过 squid 反向代理可以加速网站的访问速度，可将不同的 URL 请求分发到后台不同的 Web 服务器上，同时互联网用户只能看到反向代理服务器的地址，加强了网站的访问安全。

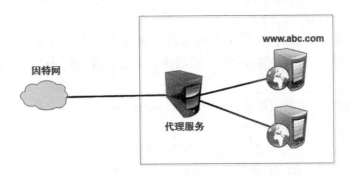

图 1.5 Squid 反向代理

Squid 反向代理加速的原理描述如下：

（1）Squid 反向代理服务器位于本地 Web 服务器和 Internet 之间，客户端请求访问 Web 服务器时，DNS 将访问的域名解析为 Squid 反向代理服务器的 IP 地址，客户端将访问 Squid 代理服务器。

（2）如果 Squid 反向代理服务器中缓存了该请求的资源，则将该请求的资源返回给客户端，否则反向代理服务器将向后台的 Web 服务器请求资源，然后将应答资源返回给客户端，同时也将该资源缓存在本地，供下一个请求者使用。

关于 Squid 反向代理更加详细的介绍及案例配置，请上课工场 APP 或官网 www.kgc.cn 观看视频。

1.5 Varnish 与 Nginx 缓存服务器

1. 高性能缓存服务器 Varnish

Varnish 是一款高性能的、开源的反向代理服务器和缓存服务器。

挪威最大的在线报纸 Verdens Gang（vg.no）使用了 3 台 Varnish 代替了原来的 12 台 Squid，性能更好！

Varnish 与 Squid 的对比如下：

（1）优点：

Varnish 具有更好的稳定性、更快的访问速度、更多的并发连接支持数，可以通过管理端口来管理缓存。

（2）缺点：

- 在高并发状态下，Varnish 消耗更多的 CPU、I/O 和内存资源。
- Varnish 进程一旦挂起、崩溃或者重启，缓存的数据会从内存中释放，此时所有的请求都会转发到后端服务器上，给后端服务器造成很大压力。

2. 轻量级缓存服务器 Nginx

Nginx 支持类似 Squid 的缓存功能，把 URL 以及相关信息当成 key，用 MD5 编码 Hash 后把数据文件保存在硬盘上。

Nginx 只能为指定的 URL 或者状态码设置过期时间，并不支持类似 Squid 的 purge 命令来手动清除指定缓存页面。可以通过第三方的 ngx_cache_purge 来清除指定的 URL 缓存。

Nginx 的缓存加速功能是由 proxy_cache 和 fastcgi_cache 两个功能模块完成的。

Nginx 缓存加速的特点如下：

- 缓存功能十分稳定，运行速度不逊于 Squid。
- 对多核 CPU 的利用率比其他的开源软件要好。
- 支持高并发请求数，能同时承受更多的访问请求。

关于 Varnish 与 Nginx 缓存更加详细的介绍及案例配置，请上课工场 APP 或官网

www.kgc.cn 观看视频。

本章总结

- 使用传统代理时，要求手动指定代理服务器的地址、端口等信息。对于浏览器来说，DNS 请求也将发送给代理服务器。
- 使用透明代理时，需要结合客户机的默认路由、网关的 REDIRECT 策略来实现，因此不适用于 Internet 环境。对于客户机浏览器来说，DNS 请求优先发送给指定的 DNS 服务器。
- 构建透明代理服务时，http_port 行需添加 transparent 监听选项，另外还需设置防火墙的 REDIRECT 策略。
- Squid 服务的访问控制主要通过 acl、http_access 配置项来设置，前者用来定义控制条件，后者决定允许和拒绝。
- Squid 日志分析工具 Sarg 采用 HTML 格式，详细列出每一位用户访问 Internet 的站点信息、时间占用信息、排名、连接次数和访问量等。
- 通过 squid 反向代理可以加速网站的访问速度，可将不同的 URL 请求分发到后台不同的 Web 服务器上，同时互联网用户只能看到反向代理服务器的地址，加强了网站的访问安全。
- Varnish 是一款高性能的、开源的反向代理服务器和缓存服务器，Nginx 是一款轻量级的缓存服务器。

本章作业

1．简述传统代理与透明代理各自的特点、适用环境。

2．在 Squid 服务器的配置项中，maximum_object_size 与 reply_body_max_size 的作用分别是什么？

3．构建 Squid 代理服务器时，IP 转发（ip_forward）功能是否必须打开？为什么？

4．为 Squid 服务设置 ACL 访问控制策略，允许 192.168.1.0/24 网段中的每台主机在白天（7:00～19:00）使用代理服务，但并发连接数不能超过 10。

5．基于 CentOS 7.3 部署 Squid 3.5.24。

6．上课工场 APP 或官网 kgc.cn 观看视频，完成 Varnish 与 Nginx 配置实验。

7．用课工场 APP 扫一扫，完成在线测试，快来挑战吧！

随手笔记

第 2 章

高性能内存对象缓存 Memcached

技能目标

- 理解 Memcached 核心概念
- 会进行 Memcached 相关部署操作
- 会进行 Memcached 主主复制操作
- 会进行 Memcached 服务高可用配置

本章导读

  Memcached 是一套开源的高性能分布式内存对象缓存系统,它将所有的数据都存储在内存中,因为在内存中会统一维护一张巨大的 Hash 表,所以支持任意存储类型的数据。很多网站使用 Memcached 以提高网站的访问速度,尤其是对于大型的需要频繁访问数据的网站。

  Memcached 主主复制是指在任意一台 Memcached 服务器修改数据都会被同步到另外一台,可以使用 keepalived 提供高可用架构。

知识服务

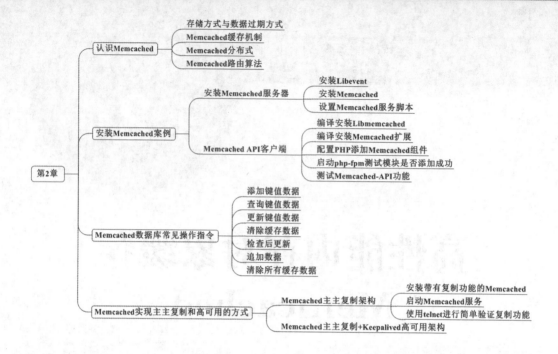

2.1 认识 Memcached

Memcached 是一套开源的高性能分布式内存对象缓存系统，它将所有的数据都存储在内存中，因为在内存中会统一维护一张巨大的 Hash 表，所以支持任意存储类型的数据。很多网站通过使用 Memcached 提高网站的访问速度，尤其是对于大型的需要频繁访问数据的网站。

Memcached 是典型的 C/S 架构，因此需要安装 Memcached 服务端与 Memcached API 客户端。Memcached 服务端是用 C 语言编写的，而 Memcached API 客户端可以用任何语言来编写，如 PHP、Python、Perl 等，并通过 Memcached 协议与 Memcached 服务端进行通信。常用典型架构如图 2.1 所示。

图 2.1 Memcached 常用架构

当 Web 客户端发送请求到 Web 服务器的应用程序时，应用程序会通过调用 Memcached API 客户端程序库接口去连接 Memcached 服务器，进而查询数据。如果此时 Web 客户端所请求的数据已经在 Memcached 服务端中缓存，则 Memcached 服务端会将数据返回给 Web 客户端；如果数据不存在，则会将 Web 客户端请求发送至 MySQL 数据库，由数据库将请求的数据返回给 Memcached 以及 Web 客户端，与此同时 Memcached 服务器也会将数据进行保存，以方便用户下次请求使用。

1. 存储方式与数据过期方式

Memcached 具有独特的存储方式和数据过期方式。

（1）数据存储方式：Slab Allocation

Slab Allocation 即按组分配内存，每次先分配一个 Slab，相当于一个大小为 1MB 的页，然后在 1MB 的空间里根据数据划分大小相同的 Chunk，如图 2.2 所示。该方法可以有效解决内存碎片问题，但可能会对内存空间有所浪费。

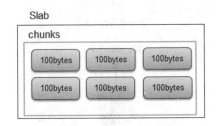

图 2.2　Slab Allocation

（2）数据过期方式：LRU、Laxzy Expiration

LRU 是指追加的数据空间不足时，会根据 LRU 的情况淘汰最近最少使用的记录。Laxzy Expiration 即惰性过期，是指使用 get 时查看记录时间，从而检查记录是否已经过期。

2. Memcached 缓存机制

缓存是常驻在内存的数据，能够快速进行读取。而 Memcached 就是这样一款非常出色的缓存软件，当程序写入缓存数据请求时，Memcached 的 API 接口将 Key 输入路由算法模块路由到集群中一台服务器，之后由 API 接口与服务器进行通信，完成一次分布式缓存写入，如图 2.3 所示。

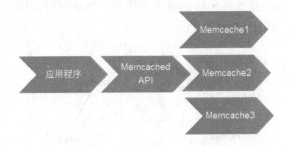

图 2.3　Memcached 缓存机制

3. Memcached 分布式

Memcached 分布式部署主要依赖于 Memcached 的客户端来实现，多个 Memcached 服务器是独立的。分布式数据如何存储是由路由算法所决定的。

当数据到达客户端程序库时，客户端的算法就依据路由算法来决定保存的 Memcached 服务器。读取数据时，客户端依据保存数据时的路由算法选中和存储数据时相同的服务器来读取数据，如图 2.4 所示。

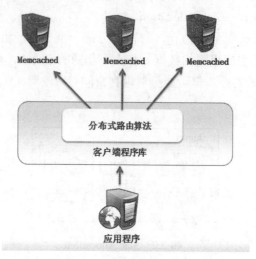

图 2.4　Memcached 分布式

4．Memcached 路由算法

（1）求余数 hash 算法

求余数 hash 算法先用 key 做 hash 运算得到一个整数，再去做 hash 算法，根据余数进行路由。这种算法适合大多数据需求，但是不适合用在动态变化的环境中，比如有大量机器添加或者删除时，会导致大量对象的存储位置失效。

（2）一致性 hash 算法

一致性 hash 算法适合在动态变化的环境中使用。原理是按照 hash 算法把对应的 key 通过一定的 hash 算法处理后，映射形成一个首尾相接的闭合循环，然后通过使用与对象存储一样的 hash 算法将机器也映射到环中，按顺时针方向将所有对象存储到离自己最近的机器中，如图 2.5 所示。

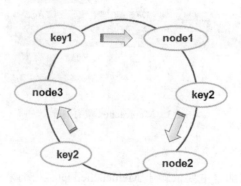

图 2.5　一致性 hash 算法

2.2 安装 Memcached 案例

本案例使用两台 CentOS7.3 系统完成，一台是 Memcached 服务器，另一台是基于 LAMP 架构进行 Memcached 扩展的 Memcached API 客户端，可以根据企业需求进行架构调整。案例环境如表 2-1 所示。

表 2-1 案例环境

名称	IP 地址	角色	主要软件包
Memcached	192.168.1.105	Memcached 服务器	libevent-1.4.12-stable.tar.gz memcached-1.4.31.tar.gz
Memcached API	192.168.1.102	Memcached API 客户端	httpd-2.4.25.tar.gz php-5.6.30.tar.gz libmemcached-1.0.18.tar.gz memcached-2.2.0.tgz

2.2.1 安装 Memcached 服务器

1. 安装 Libevent

Libevent 是一款跨平台的事件处理接口的封装，可以兼容多个操作系统的事件访问。Memcached 的安装依赖于 Libevent，因此需要先完成 Libevent 的安装。

```
[root@memcached ~]# wget https://github.com/downloads/libevent/libevent/libevent-1.4.14b-stable.tar.gz
[root@memcached ~]# tar xzvf libevent-1.4.14b-stable.tar.gz
[root@memcached libevent-1.4.14b-stable]# ./configure --prefix=/usr/local/libevent
[root@memcached libevent-1.4.14b-stable]# make && make install
```

到此 Libevent 安装完毕，接下来就可以开始安装 Memcached。

2. 安装 Memcached

采用源码的方式进行 Memcached 的编译安装，安装时需要指定 Libevent 的安装路径。

```
[root@memcached ~]# wget http://www.memcached.org/files/memcached-1.4.31.tar.gz
[root@memcached ~]# tar xzvf memcached-1.4.31.tar.gz
[root@memcached ~]# cd memcached-1.4.31/
[root@memcached memcached-1.4.31]# ./configure --prefix=/usr/local/memcached  --with-libevent=/usr/local/libevent
[root@memcached memcached-1.4.31]#make && make install
```

3. 设置 Memcached 服务脚本

Memcached 服务器安装完成后，可以使用安装目录下的 bin/memcached 来启动服务，但是为了更加方便地管理 Memcached，还是编写脚本来管理 Memcached 服务。

```bash
[root@memcached ~]# vi /usr/local/memcached/memcached_service.sh
#!/bin/bash
CMD="/usr/local/memcached/bin/memcached"
start(){
 $CMD -d -m 128 -u root
}
stop(){
 killall memcached;
}

ACTION=$1
 case $ACTION in
 'start')
     start;;
 'stop')
     stop;;
 'restart')
     stop
     sleep 2
     start;;
 *)
     echo 'Usage:{start|stop|restart}'
esac
```

其中启动 Memcached 时，-d 选项的作用是以守护进程的方式运行 Memcached 服务；-m 是为 Memcached 分配 128MB 的内存，应根据企业需要进行调整；-u 指定运行的用户账号。

之后设置脚本权限，启动 Memcached 服务。

```
[root@memcached ~]# chmod 755 /usr/local/memcached/memcached_service.sh
[root@memcached ~]# /usr/local/memcached/memcached_service.sh start
```

服务启动之后，监听 11211/tcp 端口。

```
[root@memcached ~]# netstat -antp |grep memcached
tcp        0      0 0.0.0.0:11211           0.0.0.0:*               LISTEN      10196/memcached
tcp6       0      0 :::11211                :::*                    LISTEN      10196/memcached
```

2.2.2 Memcached API 客户端

为了使得程序可以直接调用 Memcached 库和接口，可以使用 Memcached 扩展组件将 Memcached 添加为 PHP 的一个模块。此扩展使用了 Libmemcached 库提供的 API

与 Memcached 服务端进行交互。

1. 编译安装 Libmemcached

在编译 Memcached 扩展组件时，需要指定 Libmemcached 库的位置，所以先安装 Libmemcached 库。

```
[root@memcached-api ~]# wget https://launchpad.net/libmemcached/1.0/1.0.18/+download/
    libmemcached-1.0.18.tar.gz
[root@memcached-api ~]# tar xzvf libmemcached-1.0.18.tar.gz
[root@memcached-api ~]# cd libmemcached-1.0.18/
[root@memcached-api libmemcached-1.0.18]# ./configure --prefix=/usr/local/libmemcached --with-
    memcached=/usr/local/memcached
[root@memcached-api libmemcached-1.0.18]# make && make install
```

2. 编译安装 Memcached 扩展

然后就可以进行 PHP 的 Memcached 扩展组件安装。

```
[root@memcached-api ~]# wget http://pecl.php.net/get/memcached-2.2.0.tgz
[root@memcached-api ~]# tar xzvf memcached-2.2.0.tgz
[root@memcached-api ~]# cd memcached-2.2.0/
```

注意配置 Memcached API 时，memcached-2.2.0.tgz 源码包中默认没有 configure 配置脚本，需要使用 PHP 的 phpize 脚本生成配置脚本 configure。

```
[root@memcached-api memcached-2.2.0]# /usr/local/php/bin/phpize
Configuring for:
PHP Api Version:        20131106
Zend Module Api No:     20131226
Zend Extension Api No:  220131226
[root@memcached-api memcached-2.2.0]# ./configure --enable-memcached --with-php-config=/usr/
    local/php/bin/php-config --with-libmemcached-dir=/usr/local/libmemcached --disable-memcached-sasl
```

> **注意**
>
> 　　配置时使用 --disable-memcached-sasl 选项关闭 Memcached 的 SASL 认证功能，否则会报错。

```
[root@memcached-api memcached-2.2.0]# make
[root@memcached-api memcached-2.2.0]#make test
[root@memcached-api memcached-2.2.0]# make install
Installing shared extensions: /usr/local/php/lib/php/extensions/no-debug-zts-20131226/
                                                              //共享组件的位置
```

3. 配置 PHP 添加 Memcached 组件

编辑 PHP 配置文件 php.ini，添加 Memcached 组件。

```
[root@memcached-api ~]# cd /usr/local/php/
[root@memcached-api php]# vi etc/php.ini
添加如下内容
extension_dir = "/usr/local/php/lib/php/extensions/no-debug-zts-20131226/"
extension=memcached.so
```

4. 启动 php-fpm 测试模块是否添加成功

```
[root@memcached-api php]# cp etc/php-fpm.conf.default etc/php-fpm.conf
[root@memcached-api php]# ./sbin/php-fpm -c etc/php.ini -y etc/php-fpm.conf
[root@memcached-api php]# ps aux |grep php
root      83801  0.0  0.1 159340 3780 ?      Ss   08:51   0:00 php-fpm: master process (etc/php-fpm.conf)
nobody    83802  0.0  0.1 161424 3700 ?      S    08:51   0:00 php-fpm: pool www
nobody    83803  0.0  0.1 161424 3700 ?      S    08:51   0:00 php-fpm: pool www
root      83813  0.0  0.0 112648  964 pts/1  S+   08:51   0:00 grep --color=auto php
```

可以通过 phpinfo()，查看是否已经添加 Memcached 扩展模块。

```
[root@memcached ~]# vi /usr/local/apache/htdocs/index.php
<?php
phpinfo();
?>
```

使用浏览器进行访问，结果如图 2.6 所示，已经添加成功。

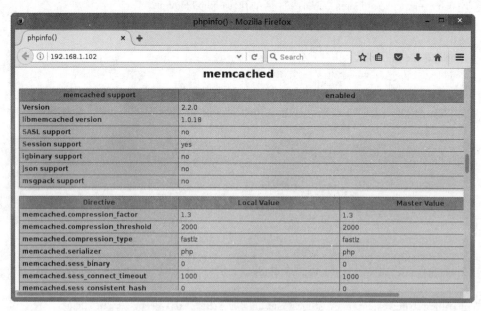

图 2.6　phpinfo 信息

5. 测试 Memcached API 功能

通过编写简单的 PHP 测试代码调用 Memcache 程序接口，来测试是否与

Memcached 服务器协同工作，代码如下：

```
[root@memcached-api ~]# vi /usr/local/apache/htdocs/test.php
<?php
$memcache = new Memcached();
$memcache->addServer('192.168.1.105', 11211);
$memcache->set('key', 'Memcache test successful!', 0, 60);
$result = $memcache->get('key');
unset($memcache);
echo $result;
?>
```

此段代码的作用是在客户端连接 Memcached 服务器，设置名为 'key' 的键的值为 'Memcache test successful!'，并读取显示。显示成功，则表示服务器与客户端协同工作正常。使用浏览器进行访问，测试结果如图 2.7 所示。

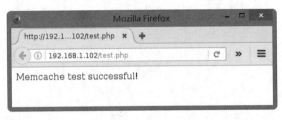

图 2.7　测试页面

2.3　Memcached 数据库操作与管理

Memcached 协议简单，可直接使用 telnet 连接 Memcached 的 11211 端口，对 Memcached 数据库进行操作与管理。

```
[root@memcached ~]# telnet 127.0.0.1 11211
Trying 127.0.0.1...
Connected to 127.0.0.1.
Escape character is '^]'.
// 输入操作指令
```

操作命令格式：

`<command name><key><flags><exptime><bytes><data block>`

1．常见操作指令

（1）添加一条键值数据

```
add username 0 0 7
example
STORED
```

其中 add username 0 0 7 表示键值名为 username，标记位表示自定义信息为 0，过期时间为 0（永不过期，单位为秒），字节数为 7。example 为键值，注意输入长度为 7 字节，与设定值相符合。

（2）查询键值数据

```
get username
VALUE username 0 7
example
END
gets username
VALUE username 0 7 4
example
END
```

其中 get 后跟键值名，如果检查最近是否更新，可以使用 gets，最后一位显示的是更新因子，每更新一次更新因子数会加 1。

（3）更新一条键值数据

```
set username 0 0 10
everything
STORED
get username
VALUE username 0 10
everything
END
```

其中 set 后跟需要更新的键值名、标记位、过期时间、字节数。如果键值名不存在，set 相当于 add。如果仅仅是想单纯地更新没有添加的功能，使用 replace。此时更新的键值名必须存在，如果键值名不存在，就会报 NOT_STORED 的错误。

```
replace username 0 0 7
lodging
STORED
gets username
VALUE username 0 7 6
lodging
END
replace username1 0 0 7
example
NOT_STORED
```

（4）清除一条缓存数据

```
delete username
DELETED
get username
END
```

使用 delete 删除一条键值为 username 的缓存数据，使用 get 查看发现没有内容存在。

（5）检查后更新

```
gets username
VALUE username 0 7 7
example
END
cas username 0 0 7 1
lodging
EXISTS
cas username 0 0 7 7
lodging
STORED
gets username
VALUE username 0 7 8
lodging
END
```

如果 cas 的最后一个更新因子数与 gets 返回的更新因子数相等，则更新，否则返回 EXISTS。

（6）追加数据

```
append username 0 0 7          // 后追加 7 字节
example
STORED
get username
VALUE username 0 14
lodgingexample
END
```

在键值名 username 的原键值后追加数据使用 append。

```
prepend username 0 0 2         // 前追加 2 字节
un
STORED
get username
VALUE username 0 16
unlodgingexample
END
```

在键值名 username 的原键值前追加数据使用 prepend。

（7）清除所有缓存数据

```
flush_all
OK
```

（8）查看服务器统计信息

```
stats
    stats items              // 返回所有键值对的统计信息
```

```
stats  cachedump 1 0      // 返回指定存储空间的键值对
stats slabs               // 显示各个 slab 的信息，包括 chunk 的大小、数目、使用情况等
stats sizes               // 输出所有 item 的大小和个数
stats reset               // 清空统计数据
```

2.4　Memcached 实现主主复制和高可用的方式

Memcached 主主复制是指在任意一台 Memcached 服务器修改数据都会被同步到另外一台，但是 Memcached API 客户端是无法判断连接到哪一台 Memcached 服务器的，所以需要设置 VIP 地址，提供给 Memcached API 客户端进行连接。可以使用 keepalived 产生的 VIP 地址连接主 Memcached 服务器，并且提供高可用架构。

本案例使用两台 Memcached 服务器来完成，实验环境如表 2-2 所示。

表 2-2　案例环境

名称	IP 地址	操作系统	主要软件包
Memcached1	192.168.1.100	Centos 7.3	libevent-1.4.12-stable.tar.gz memcached-1.2.8-repcached-2.2.tar.gz keepalived-1.2.13-8.el7.x86_64
Memcached2	192.168.1.105	Centos 7.3	libevent-1.4.12-stable.tar.gz memcached-1.2.8-repcached-2.2.tar.gz keepalived-1.2.13-8.el7.x86_64

2.4.1　Memcached 主主复制架构

Memcached 的复制功能支持多个 Memcached 之间进行相互复制（双向复制，主备都是可读可写的），可以解决 Memcached 的容灾问题。

要使用 Memcached 复制架构，需要重新下载支持复制功能的 Memcached 安装包。
http://downloads.sourceforge.net/repcached/memcached-1.2.8-repcached-2.2.tar.gz
安装过程与之前安装的 Memcached 方法相同，下面简略描述一下。

1. 安装带有复制功能的 Memcached

安装完成 Libevent 之后，将下载的 memcached-1.2.8-repcached-2.2.tar.gz 进行解压，然后完成编译安装。

```
[root@memcached1 ~]# cd memcached-1.2.8-repcached-2.2/
[root@memcached1 memcached-1.2.8-repcached-2.2]# ./configure --prefix=/usr/local/
    memcached_replication --enable-replication  --with-libevent=/usr/local/libevent
[root@memcached1 memcached-1.2.8-repcached-2.2]# make && make install
```

2. 启动 Memcached 服务

支持复制功能的 Memcached 安装完成之后，需要将编译安装的 libevent-1.4.so.2

模块复制到 /usr/lib64 目录下，否则在启动带有复制功能的 Memcached 服务时会报错。

[root@memcached1 ~]# ln -s /usr/local/libevent/lib/libevent-1.4.so.2 /usr/lib64/libevent-1.4.so.2

启动服务时，使用 -x 选项指向对端。

[root@memcached1 ~]# /usr/local/memcached_replication/bin/memcached -d -u root -m 128 -x 192.168.1.105
[root@memcached1 ~]# netstat -antp |grep memcached
tcp   0   0 0.0.0.0:11211   0.0.0.0:*   LISTEN   8163/memcached
tcp   0   0 0.0.0.0:11212   0.0.0.0:*   LISTEN   8163/memcached
tcp6   0   0 :::11211   :::*   LISTEN   8163/memcached

同样启动 Memcached2 服务器，注意启动 Memcached 服务时指向对端。

3. 使用 telnet 进行简单验证复制功能

（1）在 Memcached1 上插入一条具有特点的键值

[root@memcached1 ~]# telnet 192.168.1.100 11211
Trying 192.168.1.100...
Connected to 192.168.1.100.
Escape character is '^]'.
set username 0 0 8
20170226
STORED
get username
VALUE username 0 8
20170226
END
quit
Connection closed by foreign host.

（2）在 Memcached2 上查看刚刚插入的键值

[root@memcached2 ~]# telnet 192.168.1.105 11211
Trying 192.168.1.105...
Connected to 192.168.1.105.
Escape character is '^]'.
get username
VALUE username 0 8
20170226
END
get username2
END
quit
Connection closed by foreign host.

同理，在 Memcached2 上插入的数据，在 Memcached1 上也可以查看到。这就是 Memcached 的主主复制。

2.4.2　Memcached 主主复制 +Keepalived 高可用架构

因为 Memcached 主主复制这种架构，在程序连接时不知道应该连接哪个主服务器，所以需要在前端加 VIP 地址，实现高可用架构。这里用 Keepalived 实现，因而 Keepalived 的作用是用来检测 Memcached 服务器的状态是否正常，如图 2.8 所示。

Keepalived 不断检测 Memcached 主服务器的 11211 端口，如果检测到 Memcached 服务发生宕机或者死机等情况，就会将 VIP 从主服务器移至从服务器，从而实现 Memcached 的高可用性。

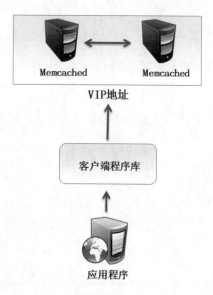

图 2.8　Memcached 高可用架构

1. 安装配置 keepalived

[root@memcached1 ~]# yum install keepalived

（1）配置主 keepalived

[root@memcached1 ~]# vi /etc/keepalived/keepalived.conf
! Configuration File for keepalived

global_defs {
  notification_email {
    admin@example.com
  }
  notification_email_from Alexandre.Cassen@firewall.loc
  smtp_server 192.168.200.1
  smtp_connect_timeout 30
  router_id LVS_DEVEL   // 路由标识，主从保持一致
}

```
vrrp_sync_group cluster {
 group {
   mem_ha
 }
}
vrrp_instance mem_ha {
    state MASTER              // 主备状态均为 MASTER
    interface ens33
    virtual_router_id 51      // 虚拟路由 ID，主从相同
    priority 100              // 优先级，主的高于备
    advert_int 1
    nopreempt                 // 不主动抢占资源，只在 Master 或者高优先级服务器上进行设置

    authentication {
      auth_type PASS
      auth_pass 1111
    }
    virtual_ipaddress {       // 定义 VIP 地址
      192.168.1.200
    }
}

virtual_server 192.168.1.200 11211{    //VIP 故障检测
    delay_loop 6
    persistence_timeout 20
    protocol TCP
    sorry_server 192.168.1.100 11211   // 对端
    real_server 192.168.1.105 11211 {  // 本机
      weight 3
      notify_down /root/memcached.sh   // 当 memcached 宕机，停止 keepalived 服务

      TCP_CHECK {
        connect_timeout 3
        nb_get_retry 3
        delay_before_retry 3
        connect_port 11211
      }
    }
}
```

设置执行脚本如下：

```
[root@memcached1 ~]# echo "/usr/bin/systemctl stop keepalived" > memcached.sh
[root@memcached1 ~]# chmod +x memcached.sh
```

（2）配置备 keepalived

主从 keepalived 配置文件内容差不多，可以直接复制进行修改，以下只把不一样

的地方整理出来。

```
[root@memcached1 ~]# scp /etc/keepalived/keepalived.conf 192.168.1.105:/etc/keepalived/
// 省略
vrrp_instance mem_ha {
    state MASTER                    // 从也使用 MASTER
    interface ens33
    virtual_router_id 51
    priority 90                     // 优先级低
    advert_int 1
                                    // 去掉 nopreempt
    authentication {
        auth_type PASS
        auth_pass 1111
    }
// 省略
virtual_server 192.168.1.200 11211 {
    delay_loop 6
    persistence_timeout 20
    protocol TCP
    sorry_server 192.168.1.100 11211     // 对端
    real_server 192.168.1.105 11211 {    // 本机
        weight 3
        notify_down /root/memcached.sh
// 省略
```

同样设置脚本如下：

```
[root@memcached2 ~]# echo "/usr/bin/systemctl stop keepalived" > memcached.sh
[root@memcached2 ~]# chmod +x memcached.sh
```

2. 测试验证

分别启动主从的 keepalived 服务。

```
[root@memcached1 ~]# systemctl start keepalived
[root@memcached2 ~]# systemctl start keepalived
```

（1）验证主 keepalived 获取 VIP 地址

使用 ip address show 命令查看 VIP 地址（使用 ifconfig 查看不到）。

```
[root@memcached1 ~]# ip address
// 省略
2: ens33: <BROADCAST,MULTICAST,UP,LOWER_UP> mtu 1500 qdisc pfifo_fast state UP qlen 1000
    link/ether 00:0c:29:1c:8c:62 brd ff:ff:ff:ff:ff:ff
    inet 192.168.1.100/24 brd 192.168.1.255 scope global ens33
       valid_lft forever preferred_lft forever
    inet 192.168.1.200/32 scope global ens33          // 已获得 VIP 地址
```

>   valid_lft forever preferred_lft forever
>   inet6 fe80::20c:29ff:fe1c:8c62/64 scope link
>   valid_lft forever preferred_lft forever

（2）验证高可用性

关闭 Memcached1 服务器的 Memcached 服务，在 Memcached2 服务器上查看地址信息。

> [root@memcached1 ~]# killall memcached
> [root@memcached2 ~]# ip addr
> // 省略
> 2: ens33: <BROADCAST,MULTICAST,UP,LOWER_UP> mtu 1500 qdisc pfifo_fast state UP qlen 1000
>   link/ether 00:0c:29:db:af:6a brd ff:ff:ff:ff:ff:ff
>   inet 192.168.1.105/24 brd 192.168.1.255 scope global ens33
>   valid_lft forever preferred_lft forever
>   inet 192.168.1.200/32 scope global ens33   // 已获取 VIP 地址
>   valid_lft forever preferred_lft forever
>   inet6 fe80::20c:29ff:fedb:af6a/64 scope link
>   valid_lft forever preferred_lft forever

本章总结

- Memcached 是分布式内存对象缓存系统，因为所有数据都存储在内存中，从而常用于网站加速。
- Memcached 分布式实现不是在服务端实现的而是在客户端实现的。
- Memcached 可以通过 keepalived 实现 Memcached 服务的高可用性。

随手笔记

第 3 章

rsync 远程同步

技能目标

- 学会配置 rsync 备份源
- 学会使用 rsync 下行、上行异地备份
- 学会使用 rsync+inotify 实时备份

本章导读

正确、有效的备份方案是保障系统及数据安全的重要手段。在服务器中，通常会结合计划任务、Shell 脚本来执行本地备份。为了进一步提高备份的可靠性，使用异地备份也是非常有必要的。

本章将要学习 rsync（Remote Sync，远程同步）工具的使用，以实现快速、安全、高效的异地备份，如构建 Web 镜像站点。

知识服务

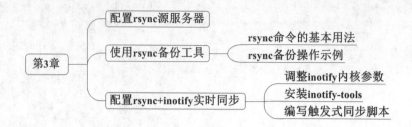

3.1 配置 rsync 源服务器

rsync 是一个开源的快速备份工具，可以在不同主机之间镜像同步整个目录树，支持增量备份，保持链接和权限，且采用优化的同步算法，在传输前执行压缩，因此非常适用于异地备份、镜像服务器等应用。

rsync 的官方站点是 http://rsync.samba.org/，由 Wayne Davison 进行维护。作为一种最常用的文件备份工具，rsync 往往是 Linux 和 UNIX 系统默认安装的基本组件之一。

```
[root@localhost ~]# rpm -q rsync
rsync-3.0.6-9.el6_4.1.x86_64
```

在远程同步任务中，负责发起 rsync 同步操作的客户机称为发起端，而负责响应来自客户机的 rsync 同步操作的服务器称为同步源。在同步过程中，同步源负责提供文档的原始位置，而发起端对该位置具有读取权限，如图 3.1 所示。

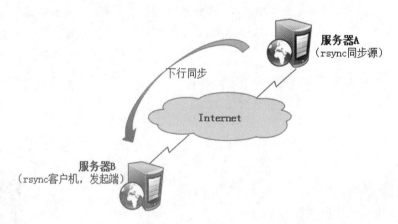

图 3.1 rsync 同步示意图

下面介绍如何配置 rsync 同步源。

rsync 作为同步源时以守护进程运行，为其他客户机提供备份源。配置 rsync 同步源需要建立配置文件 rsyncd.conf，首先创建备份账号，然后将 rsync 程序以 --daemon 选项运行。

（1）建立 /etc/rsyncd.conf 配置文件。

配置文件 rsyncd.conf 位于 /etc 目录下，需自行建立。下面将以源目录 /var/www/

html 和备份账号 backuper 为例,介绍其配置方法。

```
[root@localhost ~]# vi /etc/rsyncd.conf
uid = nobody
gid = nobody
use chroot = yes                              // 禁锢在源目录
address = 192.168.4.200                       // 监听地址
port 873                                      // 监听端口
log file = /var/log/rsyncd.log                // 日志文件位置
pid file = /var/run/rsyncd.pid                // 存放进程 ID 的文件位置
hosts allow = 192.168.4.0/24                  // 允许访问的客户机地址
[wwwroot]                                     // 共享模块名称
    path = /var/www/html                      // 源目录的实际路径
    comment = Document Root of www1.kgc.cn
    read only = yes                           // 是否为只读
    dont compress = *.gz *.bz2 *.tgz *.zip *.rar *.z   // 同步时不再压缩的文件类型
    auth users = backuper                     // 授权账户
    secrets file = /etc/rsyncd_users.db       // 存放账户信息的数据文件
```

基于安全性考虑,对于 rsync 的同步源最好仅允许以只读方式做同步。另外,同步可以采用匿名的方式,只要将其中的 auth users 和 secrets file 配置记录去掉就可以了。

(2)为备份账户创建数据文件。

根据上一步的设置,创建账号数据文件,添加一行用户记录,以冒号分隔,用户名称为 backuper,密码为 pwd123。由于账号信息采用明文存放,因此应调整文件权限,避免账号信息泄露。

```
[root@localhost ~]# vi /etc/rsyncd_users.db
backuper:pwd123                               // 无须建立同名系统用户
[root@localhost ~]# chmod 600 /etc/rsyncd_users.db
```

备份用户 backuper 需要对源目录 /var/www/html 有相应的读取权限。实际上,只要 other 组有读取权限,则备份用户 backuper 和运行用户 nobody 也就有读取权限了。

```
[root@localhost ~]# ls -ld /var/www/html/
drwxr-xr-x. 2 root root 4096 8月   2 2016 /var/www/html/
```

(3)启动 rsync 服务程序,运行参数为 --daemon。

完成上述操作以后,执行 rsync --daemon 命令就可以启动 rsync 服务,以独立监听服务的方式运行。若要关闭 rsync 服务,可以采取 kill 进程的方式,如 kill $(cat /var/run/rsyncd.pid)。

```
[root@localhost ~]# rsync --daemon            // 启动 rsync 服务
[root@localhost ~]# netstat -anpt | grep rsync
tcp     0    0 192.168.4.200:873    0.0.0.0:*    LISTEN    21182/rsync
```

3.2 使用 rsync 备份工具

有了同步源服务器之后，就可以使用 rsync 工具来执行远程同步了。本节介绍的备份操作均在客户机（发起端）执行，如服务器 B（图 3.1）。实际上，同步源与发起端可以是同一台主机（当然这种情况并不常见），其效果相当于本地备份而不是异地备份。

1. rsync 命令的基本用法

绝大多数的备份程序要求指定原始位置和目标位置，rsync 命令也一样。最简单的 rsync 用法类似于 cp 命令。例如，可以将文件 /etc/fstab 和目录 /boot/grub 同步备份到 /opt 目录下，其中，"-r" 选项表示递归整个目录树，"-l" 选项用来备份链接文件。

```
[root@localhost ~]# rsync /etc/fstab /opt
[root@localhost ~]# rsync -rl /etc/fstab /boot/grub /opt
```

（1）命令格式及常用备份选项

从以上操作可以看出，备份的基本格式为"rsync [选项] 原始位置 目标位置"，其中常用的一些命令选项如下所示，应根据实际需求做出选择（如 -avz）。

- -r：递归模式，包含目录及子目录中的所有文件。
- -l：对于符号链接文件仍然复制为符号链接文件。
- -v：显示同步过程的详细（verbose）信息。
- -a：归档模式，保留文件的权限、属性等信息，等同于组合选项 -rlptgoD。
- -z：在传输文件时进行压缩（compress）。
- -p：保留文件的权限标记。
- -t：保留文件的时间标记。
- -g：保留文件的属组标记（仅超级用户使用）。
- -o：保留文件的属主标记（仅超级用户使用）。
- -H：保留硬连接文件。
- -A：保留 ACL 属性信息。
- -D：保留设备文件及其他特殊文件。
- --delete：删除目标位置有而原始位置没有的文件。
- --checksum：根据校验和（而不是文件大小、修改时间）来决定是否跳过文件。

（2）配置源的表示方法

在执行远程同步任务时，rsync 命令需要指定同步源服务器中的资源位置。rsync 同步源的资源表示方式为"用户名 @ 主机地址 :: 共享模块名"或者"rsync:// 用户名 @ 主机地址 / 共享模块名"，前者为两个冒号分隔形式，后者为 URL 地址形式。例如，执行以下操作将访问 rsync 同步源，并下载到本地 /root 目录下进行备份。

```
[root@localhost ~]# rsync -avz backuper@192.168.4.200::wwwroot /root
```

或者

```
[root@localhost ~]# rsync -avz rsync://backuper@192.168.4.200/wwwroot /root
```

2．rsync 备份操作示例

执行以下操作将访问源服务器中的 wwwroot 共享模块，并下载到本地的 /myweb 目录下。

```
[root@localhost ~]# mkdir /myweb
[root@localhost ~]# rsync -avzH --delete backuper@192.168.4.200::wwwroot /myweb
Password:                           // 验证 backuper 用户的密码
receiving incremental file list
./
index.html
index.php
sent 102 bytes  received 243 bytes  76.67 bytes/sec
total size is 35  speedup is 0.10
[root@localhost ~]# ls /myweb        // 确认同步结果
index.html  index.php
```

实际生产环境中的备份工作通常是按计划重复执行的。例如，每天晚上 22:30 对服务器的网站目录做一次同步，定期任务可以交给 crond 服务来完成。

为了在同步过程中不用输入密码，需要创建一个密码文件，保存 backuper 用户的密码，如 /etc/server.pass。在执行 rsync 同步时使用选项 --password-file=/etc/server.pass 指定即可。

```
[root@localhost ~]# cat /etc/server.pass
pwd123
[root@localhost ~]# chmod 600 /etc/server.pass
[root@localhost ~]# crontab -e
30 22 * * * /usr/bin/rsync -az --delete --password-file=/etc/server.pass backuper@192.168.4.200::
    wwwroot /myweb                  // 每天 22:30 执行脚本
[root@localhost ~]# service crond restart
[root@localhost ~]# chkconfig crond on
```

3.3　配置 rsync+inotify 实时同步

Linux 内核从 2.6.13 版本开始提供了 inotify 通知接口，用来监控文件系统的各种变化情况，如文件存取、删除、移动、修改等。利用这一机制，可以非常方便地实现文件异动告警、增量备份，并针对目录或文件的变化及时作出响应。

将 rsync 工具与 inotify 机制相结合，可以实现触发式备份（实时同步）——只要原始位置的文档发生变化，就立即启动增量备份操作，如图 3.2 所示，否则处于静默等待状态。这样，就避免了按固定周期备份时存在的延迟性、周期过密等问题。

正因为 inotify 通知机制由 Linux 内核提供，因此主要做本机监控，在触发式备份中应用时更适合上行同步。下面依次介绍其配置过程。

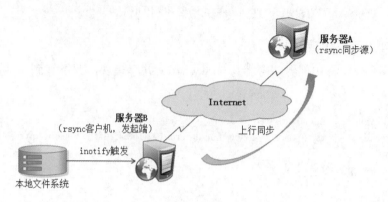

图 3.2　inotify+rsync 触发式上行同步

1. 调整 inotify 内核参数

在 Linux 内核中，默认的 inotify 机制提供了三个调控参数：max_queue_events、max_user_instances 和 max_user_watches，分别表示监控事件队列（16 384）、最多监控实例数（128）和每个实例最多监控文件数（8192）。

```
[root@localhost ~]# cat /proc/sys/fs/inotify/max_queued_events
16384
[root@localhost ~]# cat /proc/sys/fs/inotify/max_user_instances
128
[root@localhost ~]# cat /proc/sys/fs/inotify/max_user_watches
8192
```

当要监控的目录、文件数量较多或者变化较频繁时，建议加大这三个参数的值。例如，可以直接修改 /etc/sysctl.conf 配置文件，将管理队列设为 32 768，实例数设为 1024，监控数设为 1048 576（建议大于监控目标的总文件数）。

```
[root@localhost ~]# vi /etc/sysctl.conf
……                          // 省略部分信息
fs.inotify.max_queued_events = 16384
fs.inotify.max_user_instances = 1024
fs.inotify.max_user_watches = 1048576
[root@localhost ~]# sysctl -p
```

2. 安装 inotify-tools

使用 inotify 机制还需要安装 inotify-tools，以便提供 inotifywait 和 inotifywatch 辅助工具程序，用来监控和汇总改动情况。inotify-tools 可以从网站 http://inotify-tools.sourceforge.net/ 下载，版本为 3.14。

```
[root@localhost ~]# tar zxf inotify-tools-3.14.tar.gz
[root@localhost ~]# cd inotify-tools-3.14
[root@localhost inotify-tools-3.14]# ./configure
```

```
[root@localhost inotify-tools-3.14]# make
[root@localhost inotify-tools-3.14]# make install
```

以监控网站目录 /var/www/html 为例,可以先执行 inotifywait 命令,然后在另一个终端向 /var/www/html 目录下添加、移动文件,跟踪屏幕输出结果。其中,选项 -e 用来指定要监控哪些事件,选项 -m 表示持续监控,选项 -r 表示递归整个目录,选项 -q 简化输出信息。

```
[root@localhost ~]# inotifywait -mrq -e modify,create,move,delete /var/www/html
Setting up watches.  Beware: since -r was given, this may take a while!
Watches established.
/var/www/html/ CREATE index.php              // 创建 index.php 文件
/var/www/html/ MODIFY index.php              // 修改 index.php 文件
/var/www/html/ MOVED_FROM index.php          // 重命名 index.php 文件
/var/www/html/ MOVED_TO test.php             // 改名为 test.php 文件
……                                           // 省略部分信息
```

inotifywait 可监控 modify(修改)、create(创建)、move(移动)、delete(删除)、attrib(属性更改)等各种事件,一旦有变动就立即输出结果;inotifywatch 可用来收集文件系统变动情况,并在运行结束后输出汇总的变化情况。关于这两个命令的详细用法可以参考其 man 手册页,配置触发备份任务时只要用到 inotifywait 就可以了。

3. 编写触发式同步脚本

使用 inotifywait 输出的监控结果中,每行记录中依次包括目录、事件、文件,据此可以识别变动情况。为了简单,只要检测到变动时执行 rsync 上行同步操作即可。需要注意的是,当更新较频繁时,应避免并发执行 rsync 备份——若 rsync 进程已经存在,则忽略本次同步,或者根据 rsync 进程数量(取决于实际任务)来决定是否同步。

```
[root@localhost ~]# vi /opt/inotify_rsync.sh
#!/bin/bash
INOTIFY_CMD="inotifywait -mrq -e modify,create,attrib,move,delete /var/www/html/"
RSYNC_CMD="rsync -azH --delete --password-file=/etc/server.pass /var/www/html rput@
    192.168.4.200:/var/www/html"
$INOTIFY_CMD | while read DIRECTORY EVENT FILE
do
  if [ $(pgrep rsync | wc -l) -le 0 ] ; then
     $RSYNC_CMD
  fi
done
[root@localhost ~]# chmod +x /opt/inotify_rsync.sh
[root@localhost ~]# echo '/opt/inotify_rsync.sh' >> /etc/rc.local
```

上述脚本用来检测本机 /var/www/html 目录的变动情况,一旦有更新触发 rsync 同步操作,上传备份至服务器 192.168.4.200 的 /var/www/html 目录下。

触发式上行同步的验证过程如下:

(1)在本机运行 /opt/inotify_rsync.sh 脚本程序。

(2)切换到本机的 /var/www/html 目录,执行增加、删除、修改文件等操作。

（3）查看服务器中的 /var/www/html 目录下的变化情况。

本章总结

- 主控 rsync 备份操作的客户机称为发起端，负责相应备份操作的服务器称为同步源。
- 当同步源为 rsync 服务模式时，访问地址采用 rsync:// 或双冒号分隔形式。
- Linux 内核的 inotify 机制为文件系统的增加、删除、移动、修改等事件提供通知，便于实现触发式管理任务。安装 inotify-tools 软件包可以获得 inotifywait 监控程序。

本章作业

1. 使用 rsync 同步工具时，"发起端""同步源"的含义是什么？
2. 在本地新建临时目录 syncdir，使用 rsync 对远程主机的 /usr/share/doc 目录做下行同步，并记录所耗费的时间；然后适当修改远程主机的 /usr/share/doc 目录结构，再次执行同步，记录所用时间。
3. 编写一个监控脚本，使用 inotifywait 工具检测目录 /opt 的变动情况，包括删除、新增、修改、移动事件。当出现删除操作时，记录被删除目录的位置和名称，并追加到日志文件 /var/log/delete.log 中。
4. 用课工场 APP 扫一扫，完成在线测试，快来挑战吧！

第4章

MFS 分布式文件系统

技能目标

- 熟悉 MFS 文件系统的组成
- 熟悉 MFS 读写数据的处理过程
- 学会搭建 MFS 文件系统
- 能够对 MFS 进行故障恢复

本章导读

几台 Web 服务器通过 NFS 共享一个存储，在业务功能上可以满足需求，但在性能与容量上，NFS 无法胜任更高的要求。MFS 即 MooseFS，可以提供容量 PB 级别的共享存储，无需昂贵的专业硬件服务器，自身就拥有冗余功能及动态扩容功能，能够保证数据的安全性。本章将介绍 MFS 分布式文件系统的原理及环境搭建。

知识服务

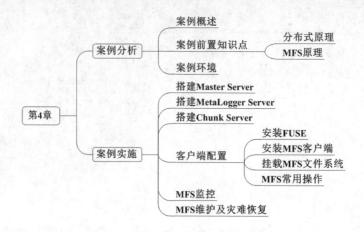

4.1 案例分析

1. 案例概述

公司之前的图片服务器采用的是 NFS，随着业务量增加，多台服务器通过 NFS 方式共享一个服务器的存储空间，使得 NFS 服务器不堪重负，经常出现超时问题。而且 NFS 存在着单点故障问题，尽管可以用 rsync 同步数据到另外一台服务器上做 NFS 服务的备份，但这对提高整个系统的性能并无帮助。基于目前的需求，我们需要对 NFS 服务器进行优化或采取别的解决方案，然而优化并不能应对日益增多的客户端的性能要求，因此选择的解决方案是采用分布式文件系统。采用分布式文件系统后，服务器之间的数据访问不再是一对多的关系，而是多对多的关系，这样可以使性能得到大幅提升。

在当前多种常用的分布式文件系统中，我们采用了 MFS（MooseFS）。MFS 正式推出是在 2008 年 5 月，它是一个具有容错功能的、高可用、可扩展的海量级分布式文件系统。MFS 把数据分散在多台服务器上，但用户看到的只是一个源。MFS 也像其他类 UNIX 文件系统一样，包含了层级结构和文件属性，可以创建特殊的文件（块设备、字符设备、管道、套接字）、符号链接和硬链接。

2. 案例前置知识点

（1）分布式原理

分布式文件系统（Distributed File System）是指文件系统管理的物理存储资源不一定直接连接在本地节点上，而是通过计算机网络与节点相连。简单来说，就是把一些分散的（分布在局域网内各个计算机上）共享文件夹，集合到一个文件夹内（虚拟共享文件夹）。对于用户来说，要访问这些共享文件夹时，只要打开这个虚拟共享文件夹，就可以看到所有链接到虚拟共享文件夹内的共享文件夹，用户感觉不到这些共享文件是分散在各个计算机上的。分布式文件系统的好处是集中访问、简化操作、数据容灾，以及提高文件的存取性能。

（2）MFS 原理

MFS 是一个具有容错性的网络分布式文件系统，它把数据分散存放在多个物理服务器上，而呈现给用户的则是一个统一的资源。

1）MFS 文件系统的组成。

MFS 文件系统的组成架构如图 4.1 所示。

- 元数据服务器（Master）：在整个体系中负责管理文件系统，维护元数据。
- 元数据日志服务器（MetaLogger）：备份 Master 服务器的变化日志文件，文件类型为 changelog_ml.*.mfs。当 Master 服务器数据丢失或者损坏时，可以从日志服务器中取得文件，进行恢复。
- 数据存储服务器（Chunk Server）：真正存储数据的服务器。存储文件时，会把文件分块保存，并在数据服务器之间进行复制。数据服务器越多，能使用的"容量"就越大，可靠性就越高，性能也就越好。
- 客户端（Client）：可以像挂载 NFS 一样挂载 MFS 文件系统，其操作是相同的。

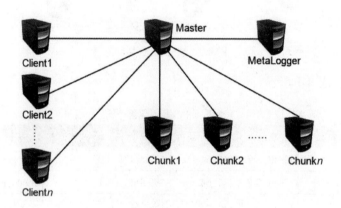

图 4.1 MFS 文件系统的组成架构

2）MFS 读取数据的处理过程。

- 客户端向元数据服务器发出读请求。
- 元数据服务器把所需数据存放的位置（Chunk Server 的 IP 地址和 Chunk 编号）告知客户端。
- 客户端向已知的 Chunk Server 请求发送数据。
- Chunk Server 向客户端发送数据。

3）MFS 写入数据的处理过程。

- 客户端向元数据服务器发送写入请求。
- 元数据服务器与 Chunk Server 进行交互（只有当所需的分块 Chunks 存在的时候才进行这个交互），但元数据服务器只在某些服务器创建新的分块 Chunks，创建成功后由 Chunk Servers 告知元数据服务器操作成功。
- 元数据服务器告知客户端，可以在哪个 Chunk Server 的哪些 Chunks 写入数据。

- 客户端向指定的 Chunk Server 写入数据。
- 该 Chunk Server 与其他 Chunk Server 进行数据同步，同步成功后 Chunk Server 告知客户端数据写入成功。
- 客户端告知元数据服务器本次写入完毕。

3. 案例环境

本案例使用六台服务器模拟搭建 MFS 文件系统，具体的拓扑如图 4.2 所示。

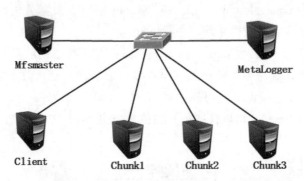

图 4.2　MFS 实验拓扑

案例环境如表 4-1 所示。

表 4-1　案例环境

主机	操作系统	IP 地址	主要软件
Master Server	CentOS 7.3 x86_64	192.168.1.100	moosefs-3.0.84-1.tar.gz
MetaLogger Server	CentOS 7.3 x86_64	192.168.1.101	moosefs-3.0.84-1.tar.gz
Chunk Server1	CentOS 7.3 x86_64	192.168.1.102	moosefs-3.0.84-1.tar.gz
Chunk Server2	CentOS 7.3 x86_64	192.168.1.103	moosefs-3.0.84-1.tar.gz
Chunk Server3	CentOS 7.3 x86_64	192.168.1.104	moosefs-3.0.84-1.tar.gz
Client	CentOS 7.3 x86_64	192.168.1.105	moosefs-3.0.84-1.tar.gz fuse-2.9.7.tar.gz

4.2　案例实施

1. 搭建 Master Server

（1）下载源码包及安装准备

分别在这几台服务器上关闭防火墙、设置地址解析记录、安装相关依赖包。

```
[root@mfsmaster ~]# systemctl stop firewalld
[root@mfsmaster ~]# systemctl disable firewalld
```

```
[root@mfsmaster ~]# vi /etc/hosts
192.168.1.100 mfsmaster
192.168.1.101 metalogger
192.168.1.102 chunk1
192.168.1.103 chunk2
192.168.1.104 chunk3
192.168.1.105 client
[root@mfsmaster ~]# yum -y install gcc zlib-devel
```

（2）创建用户

```
[root@mfsmaster ~]# useradd -s /sbin/nologin -M mfs
```

（3）安装源码包

```
[root@mfsmaster ~]# tar xf moosefs-3.0.84-1.tar.gz
[root@mfsmaster ~]# cd moosefs-3.0.84/
[root@mfsmaster moosefs-3.0.84 ]# ./configure --prefix=/usr/local/mfs --with-default-user=mfs --with-default-group=mfs --disable-mfschunkserver --disable-mfsmount
[root@mfsmaster moosefs-3.0.84]# make
[root@mfsmaster moosefs-3.0.84]# make install
```

（4）复制文件

```
[root@mfsmaster moosefs-3.0.84]# cd /usr/local/mfs/etc/mfs
[root@mfsmaster mfs]# ll
total 24
-rw-r--r--. 1 root root 4057 Feb 20 16:26 mfsexports.cfg.sample
-rw-r--r--. 1 root root 8597 Feb 20 16:26 mfsmaster.cfg.sample
-rw-r--r--. 1 root root 2175 Feb 20 16:26 mfsmetalogger.cfg.sample
-rw-r--r--. 1 root root 1052 Feb 20 16:26 mfstopology.cfg.sample
[root@mfsmaster mfs]# cp mfsmaster.cfg.sample mfsmaster.cfg
[root@mfsmaster mfs]# cp mfsexports.cfg.sample mfsexports.cfg
[root@mfsmaster mfs]# cp mfstopology.cfg.sample mfstopology.cfg
[root@mfsmaster mfs]# cd /usr/local/mfs/var/mfs
[root@mfsmaster mfs]# ll
total 4
-rw-r--r--. 1 mfs mfs 8 Feb 20 16:26 metadata.mfs.empty
[root@mfsmaster mfs]# cp metadata.mfs.empty metadata.mfs
```

（5）配置文件

需要用到的配置文件有两个：mfsmaster.cfg（主配置文件）和 mfsexports.cfg（被挂载目录及权限配置文件）。

mfsmaster.cfg 文件常用参数如下：

```
# RUNTIME OPTIONS  #
# WORKING_USER=mfs         // 运行 masterserver 的用户
# WORKING_GROUP=mfs        // 运行 masterserver 的组
# SYSLOG_IDENT=mfsmaster   // 在 syslog 中表示是 mfsmaster 产生的日志
```

```
                                            //masterserver 在 syslog 的标识，说明是由 masterserver 产生的
# LIMIT_GLIBC_MALLOC_ARENAS=4
# DISABLE_OOM_KILLER=1
# LOCK_MEMORY=0      // 是否执行 mlockall()，以避免 mfsmaster 进程溢出（默认为 0）
# NICE_LEVEL=-19     // 运行的优先级（如果可以，默认是 -19；注意：进程必须用 root 启动）
# FILE_UMASK = 027
# DATA_PATH=/usr/local/mfs/var/mfs       // 数据存放路径
# EXPORTS_FILENAME=/usr/local/mfs/etc/mfs/mfsexports.cfg
                    // 被挂接目录及其权限控制文件的存放位置
# TOPOLOGY_FILENAME=/usr/local/mfs/etc/mfs/mfstopology.cfg
# BACK_LOGS=50                //metadata 改变的 log 文件数目（默认是 50）
# METADATA_SAVE_FREQ = 1
# BACK_META_KEEP_PREVIOUS=1
# CHANGELOG_PRESERVE_SECONDS = 1800
# MISSING_LOG_CAPACITY = 100000

# COMMAND CONNECTION OPTIONS  #

# MATOML_LISTEN_HOST=*        //metalogger 监听的 IP 地址（默认是 *，代表任何 IP）
# MATOML_LISTEN_PORT=9419     //metalogger 监听的端口地址（默认是 9419）

# CHUNKSERVER CONNECTION OPTIONS #
# MATOCS_LISTEN_HOST=*        // 用于 chunkserver 连接的 IP 地址（默认是 *，代表任何 IP）
# MATOCS_LISTEN_PORT=9420     // 用于 chunkserver 连接的端口地址（默认是 9420）
# MATOCS_TIMEOUT = 10
# AUTH_CODE = mfspassword
# REMAP_BITS = 24
# REMAP_SOURCE_IP_CLASS = 192.168.1.0
# REMAP_DESTINATION_IP_CLASS = 10.0.0.0

# CHUNKSERVER WORKING OPTIONS #
#REPLICATIONS_DELAY_INIT=300           // 延迟复制的时间（默认是 300s）
# CHUNKS_LOOP_MAX_CPS = 100000
# CHUNKS_LOOP_MIN_TIME = 300           //chunks 的回环频率（默认是 300 秒）

# CHUNKS_SOFT_DEL_LIMIT = 10
# CHUNKS_HARD_DEL_LIMIT = 25
# CHUNKS_WRITE_REP_LIMIT = 2,1,1,4     // 在一个循环里复制到一个 chunkserver 的
                                       // 最大 chunk 数
# CHUNKS_READ_REP_LIMIT = 10,5,2,5     // 在一个循环里从一个 chunkserver 复制的
                                       // 最大 chunk 数
# CS_HEAVY_LOAD_THRESHOLD = 100
# CS_HEAVY_LOAD_RATIO_THRESHOLD = 5.0
# CS_HEAVY_LOAD_GRACE_PERIOD = 900
# ACCEPTABLE_DIFFERENCE = 1.0
# PRIORITY_QUEUES_LENGTH = 1000000
```

CS_MAINTENANCE_MODE_TIMEOUT = 0
CS_TEMP_MAINTENANCE_MODE_TIMEOUT = 1800

CLIENTS CONNECTION OPTIONS
MATOCL_LISTEN_HOST=*   //用于客户端挂接连接的 IP 地址（默认是 *，代表任何 IP）
MATOCL_LISTEN_PORT=9421   //用于客户端挂接连接的端口地址（默认是 9421）

CLIENTS WORKING OPTIONS
SESSION_SUSTAIN_TIME = 86400
QUOTA_DEFAULT_GRACE_PERIOD = 604800
ATIME_MODE = 0

mfsexports.cfg 文件参数格式如下：

```
# Allow everything but "meta".
*              /        rw,alldirs,admin,maproot=0:0
# Allow "meta".
*              .        rw
```

该文件的每一个条目分为三部分。

第一部分：客户端的 IP 地址。

第二部分：被挂接的目录。

第三部分：客户端拥有的权限。

地址可以指定的几种表现形式：

- *——所有的 IP 地址。
- n.n.n.n——单个 IP 地址。
- n.n.n.n/b——IP 网络地址 / 位数掩码。
- n.n.n.n/m.m.m.m——IP 网络地址 / 子网掩码。
- f.f.f.f-t.t.t.t——IP 段。

目录部分的标示如下。

- / 标识 MFS 根。
- . 表示 MFSMETA 文件系统。

权限部分的标示如下：

- ro——只读模式共享。
- rw——读写的方式共享。
- alldirs——允许挂载任何指定的子目录。
- admin——管理员权限。
- maproot——映射为 root，还是指定的用户。
- Password——指定客户端密码。

以上是对 Master Server 的 mfsmaster.cfg 和 mfsexports.cfg 配置文件的解释，对于这两个文件不需要做任何修改就可以开始工作。

（6）启动 Master Server

```
[root@mfsmaster mfs]# /usr/local/mfs/sbin/mfsmaster start
open files limit has been set to: 16384
working directory: /usr/local/mfs/var/mfs
lockfile created and locked
initializing mfsmaster modules ...
exports file has been loaded
topology file has been loaded
loading metadata ...
metadata file has been loaded
no charts data file - initializing empty charts
master <-> metaloggers module: listen on *:9419
master <-> chunkservers module: listen on *:9420
main master server module: listen on *:9421
mfsmaster daemon initialized properly


[root@mfsmaster mfs]# ps -ef | grep mfs         // 检查是否启动
mfs       5996    1  0 03:03 ?        00:00:00 /usr/local/mfs/sbin/mfsmaster start
root      5998 5938  0 03:04 pts/0    00:00:00 grep mfs
```

停止 Master Server 的命令是 /usr/local/mfs/sbin/mfsmaster -s。

2. 搭建 MetaLogger Server

```
[root@metalogger ~]# systemctl stop firewalld [root@metalogger ~]# systemctl disable firewalld
[root@metalogger ~]# yum -y install gcc zlib-devel
 [root@metalogger ~]# useradd -s /sbin/nologin -M mfs
[root@metalogger ~]# tar xf moosefs-3.0.84-1.tar.gz
[root@metalogger ~]# cd moosefs-3.0.84/
[root@metalogger moosefs-3.0.84]# ./configure --prefix=/usr/local/mfs --with-default-user=mfs --with-
        default-group=mfs --disable-mfschunkserver --disable-mfsmount
[root@metalogger moosefs-3.0.84]# make
[root@metalogger moosefs-3.0.84]# make install

[root@metalogger mfs]# cd /usr/local/mfs/etc/mfs
[root@metalogger mfs]# ll
total 24
-rw-r--r--. 1 root root 4057 Feb 20 17:42 mfsexports.cfg.sample
-rw-r--r--. 1 root root 8597 Feb 20 17:42 mfsmaster.cfg.sample
-rw-r--r--. 1 root root 2175 Feb 20 17:42 mfsmetalogger.cfg.sample
-rw-r--r--. 1 root root 1052 Feb 20 17:42 mfstopology.cfg.sample
[root@metalogger mfs]# cp mfsmetalogger.cfg.sample mfsmetalogger.cfg

[root@metalogger mfs]# vi mfsmetalogger.cfg
# RUNTIME OPTIONS  #
# WORKING_USER = mfs
# WORKING_GROUP = mfs
```

```
# SYSLOG_IDENT = mfsmetalogger
# LOCK_MEMORY = 0
# LIMIT_GLIBC_MALLOC_ARENAS = 4
# DISABLE_OOM_KILLER = 1
# NICE_LEVEL = -19
# FILE_UMASK = 027

# DATA_PATH = /usr/local/mfs/var/mfs

# BACK_LOGS = 50
# BACK_META_KEEP_PREVIOUS = 3
# META_DOWNLOAD_FREQ = 24

# MASTER CONNECTION OPTIONS #
# MASTER_RECONNECTION_DELAY = 5

# BIND_HOST = *
 MASTER_HOST = 192.168.1.100         // 修改为 Master Server 的 IP 地址
# MASTER_PORT = 9419

# MASTER_TIMEOUT = 10


[root@metalogger mfs]# /usr/local/mfs/sbin/mfsmetalogger start
open files limit has been set to: 4096
working directory: /usr/local/mfs/var/mfs
lockfile created and locked
initializing mfsmetalogger modules ...
mfsmetalogger daemon initialized properly
[root@metalogger mfs]# ps -ef | grep mfs
mfs     5895    1  0 20:31 ?        00:00:00 /usr/local/mfs/sbin/mfsmetalogger start
root    5897 1111  0 20:32 pts/0    00:00:00 grep mfs
```

停止 MetaLogger Server 的命令是 /usr/local/mfs/sbin/mfsmetalogger -s。

3. 搭建 Chunk Server

本案例中三台 Chunk Server 的搭建步骤是相同的，具体步骤如下：

```
[root@chunk1 ~]# systemctl stop firewalld
[root@chunk1 ~]# systemctl disable firewalld
[root@chunk1 ~]# yum -y install gcc zlib-devel
[root@chunk1 ~]# useradd -s /sbin/nologin -M mfs
[root@chunk1 ~]# tar xf moosefs-3.0.84-1.tar.gz
[root@chunk1 ~]# cd moosefs-3.0.84/
[root@chunk1 mfs-1.6.27]# ./configure --prefix=/usr/local/mfs  --with-default-user=mfs --with-default-
    group=mfs --disable-mfsmaster --disable-mfsmount
[root@chunk1 moosefs-3.0.84]# make
```

```
[root@chunk1 moosefs-3.0.84]# make install

[root@chunk1 moosefs-3.0.84]# cd /usr/local/mfs/etc/mfs
[root@chunk1 mfs]# ll
total 12
-rw-r--r--. 1 root root 3491 Feb 20 17:42 mfschunkserver.cfg.sample
-rw-r--r--. 1 root root 1648 Feb 20 17:42 mfshdd.cfg.sample
-rw-r--r--. 1 root root 2175 Feb 20 17:42 mfsmetalogger.cfg.sample
 [root@chunk1 mfs]# cp mfschunkserver.cfg.sample mfschunkserver.cfg
[root@chunk1 mfs]# cp mfshdd.cfg.sample  mfshdd.cfg
[root@chunk1 mfs]# vi mfschunkserver.cfg
# RUNTIME OPTIONS #
# WORKING_USER = mfs
# WORKING_GROUP = mfs
# SYSLOG_IDENT = mfschunkserver
# LOCK_MEMORY = 0
# LIMIT_GLIBC_MALLOC_ARENAS = 4
# DISABLE_OOM_KILLER = 1
# NICE_LEVEL = -19
# FILE_UMASK = 027

# DATA_PATH = /usr/local/mfs/var/mfs
# HDD_CONF_FILENAME = /usr/local/mfs/etc/mfs/mfshdd.cfg
# HDD_TEST_FREQ = 10
# HDD_LEAVE_SPACE_DEFAULT = 256MiB
# HDD_REBALANCE_UTILIZATION = 20
# HDD_ERROR_TOLERANCE_COUNT = 2
# HDD_ERROR_TOLERANCE_PERIOD = 600 # HDD_FSYNC_BEFORE_CLOSE = 0
# HDD_SPARSIFY_ON_WRITE = 1

# WORKERS_MAX = 250
# WORKERS_MAX_IDLE = 40

# MASTER CONNECTION OPTIONS #
# LABELS =


# BIND_HOST = *
 MASTER_HOST = 192.168.1.100    // 修改为 Master Server 的 IP 地址
# MASTER_PORT = 9420

# MASTER_TIMEOUT = 60
# MASTER_RECONNECTION_DELAY = 5
# AUTH_CODE = mfspassword
```

```
# CLIENTS CONNECTION OPTIONS #
# CSSERV_LISTEN_HOST = *
# CSSERV_LISTEN_PORT = 9422
```

```
[root@chunk1 mfs]# vi mfshdd.cfg
# mount points of HDD drives
#
#/mnt/hd1
#*/mnt/hd2
…
#~/mnt/hd7
/data         // 添加一行 /data，在这里 /data 是一个给 MFS 的分区，生产环境最好使用
              // 独立的分区或磁盘挂载到此目录
[root@chunk1 mfs]# mkdir /data
[root@chunk1 mfs]# chown -R mfs:mfs /data
[root@chunk1 mfs]# /usr/local/mfs/sbin/mfschunkserver start
open files limit has been set to: 16384
working directory: /usr/local/mfs/var/mfs
lockfile created and locked
setting glibc malloc arena max to 4
setting glibc malloc arena test to 4
initializing mfschunkserver modules ...
hdd space manager: path to scan: /data/
hdd space manager: start background hdd scanning (searching for available chunks)
main server module: listen on *:9422
no charts data file - initializing empty charts
mfschunkserver daemon initialized properly
 [root@chunk1 mfs]# ps -ef | grep mfs
mfs   5699    1  0 21:56 ?        00:00:00 /usr/local/mfs/sbin/mfschunkserver start
root  5725 1111  0 21:57 pts/0    00:00:00 grep mfs
```

停止 Chunk Server 的命令是 /usr/local/mfs/sbin/mfschunkserver -s。

4．客户端配置

```
[root@client ~]# systemctl stop firewalld
[root@client ~]# systemctl disable firewalld
[root@mfsmaster ~]# vi /etc/hosts
192.168.1.100 mfsmaster
192.168.1.101 metalogger
192.168.1.102 chunk1
192.168.1.103 chunk2
192.168.1.104 chunk3
192.168.1.105 client
[root@client ~]# yum -y install gcc zlib-devel
```

（1）安装 FUSE

MFS 客户端依赖于 FUSE，其源码包下载地址为 https://github.com/libfuse/libfuse/releases/download/fuse-2.9.7/fuse-2.9.7.tar.gz。

```
[root@client ~]# tar xf fuse-2.9.7.tar.gz
[root@client ~]# cd fuse-2.9.7/
[root@client fuse-2.9.2]# ./configure
[root@client fuse-2.9.2]# make
[root@client fuse-2.9.2]# make install
```

然后设置环境变量：

```
[root@client ~]# vim /etc/profile
[root@client ~]# tail -1 /etc/profile
export PKG_CONFIG_PATH=/usr/local/lib/pkgconfig:$PKG_CONFIG_PATH
[root@client ~]# source /etc/profile
```

（2）安装 MFS 客户端

```
[root@client ~]# useradd -s /sbin/nologin -M mfs
[root@client ~]# tar xf moosefs-3.0.84-1.tar.gz
[root@client ~]# cd moosefs-3.0.84/
[root@client mfs-1.6.27]# ./configure --prefix=/usr/local/mfs --with-default-user=mfs --with-default-group=mfs --disable-mfsmaster --disable-mfschunkserver --enable-mfsmount
[root@client mfs-1.6.27]# make
[root@client mfs-1.6.27]# make install
```

（3）挂载 MFS 文件系统

```
[root@client ~]# mkdir /mnt/mfs                // 创建挂接点
[root@client ~]# modprobe fuse                 // 加载 fuse 模块到内核
[root@client ~]# /usr/local/mfs/bin/mfsmount /mnt/mfs -H 192.168.1.100    // 挂载 MFS
mfsmaster accepted connection with parameters: read-write,restricted_ip,admin; root mapped to root:root
[root@client ~]# df -TH                        // 查看挂载情况
```

Filesystem	Type	Size	Used	Avail	Use%	Mounted on
/dev/mapper/VolGroup-lv_root	ext4	19G	1.3G	17G	8%	/
tmpfs	tmpfs	519M	0	519M	0%	/dev/shm
/dev/sda1	ext4	508M	35M	448M	8%	/boot
/dev/sr0	iso9660	4.5G	4.5G	0	100%	/media/cdrom
192.168.1.100:9421	f	use.mfs	88G	6.3G	82G	8%	/mnt/mfs

如果要卸载 MFS，使用命令 umount /mnt/mfs 即可。

（4）MFS 常用操作

MFS 在客户端安装完毕后，会生成 /usr/local/mfs/bin/ 目录，在这个目录下有很多命令是用户所需要的。为了方便使用这些命令，可将 /usr/local/mfs/bin 加入到环境变量中。

```
[root@client ~]# vim /etc/profile
[root@client ~]# tail -1 /etc/profile
export PATH=/usr/local/mfs/bin:$PATH
[root@client ~]# source /etc/profile
```

mfsgetgoal 命令用来查询文件被复制的份数，利用 -r 命令可以对整个目录进行递归，goal 是指文件被复制的份数。

```
[root@client ~]# mfsgetgoal -r /mnt/mfs/
/mnt/mfs/:
 directories with goal  2 :         1
```

mfssetgoal 命令用来设置文件被复制的份数，生产环境 Chunk Server 节点数量应至少大于 2，文件副本数小于等于 Chunk Server 服务器的数量。

```
[root@client ~]# mfssetgoal -r 3 /mnt/mfs/
/mnt/mfs/:
 inodes with goal changed:          1
 inodes with goal not changed:      0
 inodes with permission denied:     0
[root@client ~]# mfsgetgoal -r /mnt/mfs/
/mnt/mfs/:
 directories with goal  3 :         1
```

创建文件测试如下：

```
[root@client ~]# cd /mnt/mfs
[root@client mfs]# touch test
[root@client mfs]# mfsgetgoal test
test: 3
```

5．MFS 监控

Mfscgiserv 是用 python 编写的一个 Web 服务器，其监听端口是 9425，可以在 Master Server 上通过命令 /usr/local/mfs/sbin/mfscgiserv 来启动，用户利用浏览器就可以全面监控所有客户挂载、Chunk Server、Master Server，以及客户端的各种操作等。

在客户端上通过浏览器访问 http://192.168.1.100:9425，如图 4.3 所示。

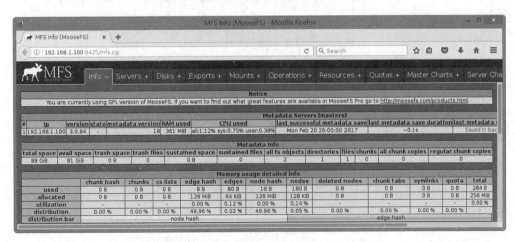

图 4.3　MFS 监控界面

其中各部分的含义如下。
- Info 部分：显示了 MFS 的基本信息。
- Servers 部分：列出现有 Chunk Server。
- Disks 部分：列出现有 Chunk Server 的硬盘信息。
- Exports 部分：列出可被挂载的目录。
- Mounts 部分：列出被挂载的目录。
- Operations 部分：显示正在执行的操作。
- Resources 部分：列出当前存储信息。
- Quotas 部分：列出当前配额信息。
- Master Charts 部分：显示 Master Server 的操作情况，如读、写、删除等操作。
- Server Charts 部分：显示 Chunk Server 的操作情况、数据传输率及系统状态。

6. MFS 维护及灾难恢复

（1）MFS 集群的启动与停止

MFS 集群启动的顺序如下：

1）启动 mfsmaster 进程。

2）启动所有的 mfschunkserver 进程。

3）启动 mfsmetalogger 进程（如果配置了 mfsmetalogger）。

4）在所有的客户端挂载 MFS 文件系统。

MFS 集群停止的顺序如下：

1）在所有的客户端卸载 MFS 文件系统。

2）用 mfschunkserver -s 命令停止 chunkserver 进程。

3）用 mfsmetalogger -s 命令停止 metalogger 进程。

4）用 mfsmaster -s 命令停止 master 进程。

（2）MFS 灾难恢复

整个 MFS 体系中，直接断电只有 Master 有可能无法启动，可以在 master 上使用命令 /usr/local/mfs/sbin/mfsmetarestore -a 修复。

MFS 元数据通常有两部分的数据，分别如下：

1）主要元数据文件 metadata.mfs，当 mfsmaster 运行时会被命名为 metadata.mfs.back。

2）元数据改变日志 changelog.*.mfs，存储了过去的 N 个小时的文件改变（N 的数值是由 BACK_LOGS 参数设置的，参数设置在 mfschunkserver.cfg 配置文件中）。

在 Master 发生故障时，可以从 MetaLogger 中恢复 Master，步骤如下：

1）安装一台 mfsmaster，利用同样的配置来配置这台 mfsmaster。

2）将 metalogger 上 /usr/local/mfs/var/mfs/ 目录下的文件复制到 mfsmaster 相应的目录中。

```
scp root@192.168.1.101:/usr/local/mfs/var/mfs/* /usr/local/mfs/var/mfs/*
```

3）利用 mfsmetarestore 命令合并元数据 changelogs。

/usr/local/mfs/sbin/mfsmetarestore -m metadata_ml.mfs.back -o metadata.mfs changelog_ml.*.mfs

如果是全新安装的 Master，恢复数据后，就要更改 metalogger 和 chunkserver 配置 MASTER_HOST 的 IP，客户端也需要重新挂载。

> **请思考**
>
> 如何保证 Master Server 的高可用性？

本章总结

- MFS 是一个具有容错性的网络分布式文件系统，它把数据分散存放在多个物理服务器上，而呈现给用户的是一个统一的资源。
- MFS 文件系统的组成包括元数据服务器（Master）、元数据日志服务器（MetaLogger）、数据存储服务器（Chunk Server）和客户端（Client）。
- 元数据服务器（Master）需要用到的配置文件有两个，分别是 mfsmaster.cfg 和 mfsexports.cfg。
- 在元数据服务器（Master）发生故障时，可以从 MetaLogger 中恢复 Master。

本章作业

1. 模拟一台数据存储服务器（Chunk Server）宕机，测试 MFS 文件系统。
2. 实时增加一台数据存储服务器（Chunk Server），验证 MFS 扩容结果。
3. 模拟 Master 发生故障，从 MetaLogger 中恢复 Master。

随手笔记

第5章

部署社交网站

技能目标

- 学会搭建 SVN 服务器
- 学会部署社交网站
- 学会对关键服务器进行双机热备
- 学会对服务器进行压力测试及性能调优

本章导读

社交网站的第一个版本部署在 LNMP 平台之上,前端为 Nginx 服务器,通过 fastcgi 协议访问后端的 PHP 服务器。为了保证数据安全,要求搭建 MySQL 数据库主从集群。社交网站项目包含用户的相册功能,允许用户上传照片,上传照片需要使用共享存储来存放。

知识服务

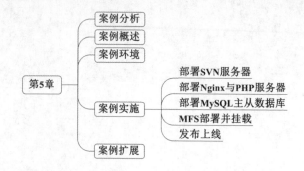

5.1 案例分析

5.1.1 案例概述

公司的社交网站采用 PHP 语言开发，为了管理 PHP 程序员开发的代码，上级领导要求搭建 SVN 服务器进行版本控制。社交网站的第一个版本部署在 LNMP 平台之上，前端为 Nginx 服务器，通过 fastcgi 协议访问后端的 PHP 服务器。为了保证数据安全，要求搭建 MySQL 数据库主从集群。

社交网站项目包含用户的相册功能，允许用户上传照片，上传照片需要使用共享存储来存放。针对共享存储可用的开源方案有很多，如 MFS、FastDFS 等。公司决定使用 MFS 分布式文件系统来实现，并将 MFS 挂载在 PHP 服务器的相关目录下。

本案例拓扑图如图 5.1 所示。

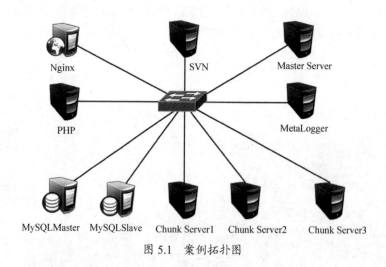

图 5.1 案例拓扑图

5.1.2 案例环境

案例环境要求见表 5-1。

表 5-1 案例环境要求

主机	操作系统	IP 地址	主要软件
Nginx	CentOS 6.5 x86_64	192.168.0.10	nginx-1.6.0.tar.gz
PHP	CentOS 6.5 x86_64	192.168.0.12	php-5.3.28.tar.gz mfs-1.6.27-5.tar.gz fuse-2.9.2.tar.gz
MySQL Master	CentOS 6.5 x86_64	192.168.0.14	mysql-5.5.22.tar.gz
MySQL Slave	CentOS 6.5 x86_64	192.168.0.15	mysql-5.5.22.tar.gz
Master Server	CentOS 6.5 x86_64	192.168.0.21	mfs-1.6.27-5.tar.gz
MetaLogger	CentOS 6.5 x86_64	192.168.0.22	mfs-1.6.27-5.tar.gz
Chunk Server1	CentOS 6.5 x86_64	192.168.0.23	mfs-1.6.27-5.tar.gz
Chunk Server2	CentOS 6.5 x86_64	192.168.0.24	mfs-1.6.27-5.tar.gz
Chunk Server3	CentOS 6.5 x86_64	192.168.0.25	mfs-1.6.27-5.tar.gz
SVN	CentOS 6.5 x86_64	192.168.0.172	

5.2 案例实施

根据公司的需求，实施过程大致分为以下步骤。

- 部署 SVN 服务器，为 PHP 程序员创建 repo 目录的访问账户，通知程序员可以导入代码。
- 部署 MySQL 主从服务器，根据 PHP 程序员的要求创建数据库与表。
- 部署 Nginx 服务器。
- 部署 PHP 服务器。
- 部署 MFS，将 MFS 文件系统挂载在前端 PHP 服务器的相关目录下。
- 通知上线部署人员可以发布上线。
- 保证数据库服务、PHP 服务、Nginx 服务依次启动，并通知测试人员开始测试，网站维护人员检查 Nginx、PHP 与数据库服务器是否正常工作。

1. 部署 SVN 服务器

（1）安装 SVN 服务。

```
[root@localhost ~]# yum –y install subversion
[root@localhost ~]# svnserve --version
svnserve, 版本 1.6.11 (r934486)
……                    // 省略部分内容
```

（2）为 PHP 程序员创建仓库目录 repo。

```
[root@localhost ~]# mkdir -p /opt/svn/repo            // 创建目录
[root@localhost ~]# svnadmin create /opt/svn/repo/    // 创建一个新的仓库
```

```
[root@localhost ~]# ls !$
ls /opt/svn/repo/
conf db format hooks locks README.txt
```

（3）调整 SVN 参数。

```
[root@localhost ~]# vim /opt/svn/repo/conf/svnserve.conf
[general]                                    // 总体配置
anon-access = none                           // 匿名用户没有任何权限
auth-access = write                          // 认证用户具有写权限
password-db = /opt/svn/repo/conf/passwd      // 用户的密码文件
authz-db = /opt/svn/repo/conf/authz          // 用户信息文件
……                                           // 省略部分内容
```

（4）启动 SVN 服务。

```
[root@localhost ~]# svnserve -d -r /opt/svn/repo/    // 关闭通过 kill PID
[root@localhost ~]# netstat -tunlp | grep svnserve
tcp    0   0 0.0.0.0:3690    0.0.0.0:*    LISTEN    2093/svnserve
```

（5）为 PHP 程序员 alpha 建立账户，配置对 repo 仓库具有读写权限，并将账户信息及仓库目录信息反馈给 PHP 程序员。

```
[root@localhost ~]# cd /opt/svn/repo
[root@localhost repo]# vim conf/passwd      // 创建账户密码文件
[users]
alpha=alphapasswd                           // 格式：用户名=用户密码
sysadmin=sysadminpasswd
[root@localhost repo]# vim /conf/authz      // 创建权限文件
[/]
sysadmin = r
alpha = r
[/webphp]
sysadmin = r                                // 运维工程师账户具有读权限，用于部署
alpha = rw                                  //alpha 用户对此目录有读写权限
[root@localhost repo]# mkdir webphp
[root@localhost repo]# svn import webphp file:///opt/svn/repo/webphp -m
" 初始化 SVN 目录 "                          // 导入 webphp
```

（6）PHP 程序员将代码导入 webphp 项目中。

可以在宿主机上安装 SVN 的 Windows 客户端，然后导入代码。

2. 部署 Nginx 服务器（192.168.0.10）与 PHP 服务器（192.168.0.12）

（1）Nginx 服务器安装。

```
[root@localhost ~]# yum -y install pcre-devel zlib-devel
[root@localhost ~]# useradd -M -s /sbin/nologin nginx
[root@localhost ~]# tar zxf nginx-1.6.0.tar.gz
[root@localhost ~]# cd nginx-1.6.0
```

```
[root@localhost nginx-1.6.0]# ./configure --prefix=/usr/local/nginx --with- http_stub_status_module
   --user=naginx --group=nginx
[root@localhost nginx-1.6.0]# make && make install
```

（2）修改 Nginx 配置。

需要配置后端 PHP 程序的 fastcgi 访问接口。

```
[root@localhost ~]# vim /usr/local/nginx/conf/nginx.conf
user  nginx;
……                            // 省略部分内容
    location / {
       root   html/webphp;
       index  index.html index.htm;
    }
location ~\.php$ {
       root   /var/www/html/webphp
       fastcgi_pass 192.168.0.12:9000;
       fastcgi_index index.php;
       include   fastcgi.conf;
    }
……                            // 省略部分内容
[root@localhost nginx-1.6.0]# ln -s /usr/local/nginx/sbin/nginx /usr/local/ sbin/
[root@localhost ~]# nginx
[root@localhost ~]# netstat -anpt | grep nginx
tcp   0   0 0.0.0.0:80   0.0.0.0:*    LISTEN   3534/nginx
```

（3）部署 PHP 服务器（192.168.0.12）。

安装 PHP，并且通过配置 php-fpm 进程监听 9000 端口来接受 Nginx 的请求。

```
[root@localhost ~]# yum -y install gd libxml2-devel libjpeg-devel libpng-devel  mysql-devel
[root@localhost ~]# useradd -M -s /sbin/nologin php
[root@localhost ~]# tar xf php-5.3.28.tar.gz
[root@localhost ~]# cd php-5.3.28
[root@localhost php-5.3.28]# cp /usr/lib64/mysql/libmysqlclient.so.16.0.0 /usr/lib/libmysqlclient.so
                  //PHP 默认去 /usr/lib/ 搜索 libmysqlclient.so
[root@localhost php-5.3.28]# ./configure --prefix=/usr/local/php --with-gd --with-zlib --with-mysql
   --with-mysqli --with-mysql-sock --with-config- file-path=/ usr/local/php --enable-mbstring
   --enable-fpm --with-jpeg-dir=/ usr/lib
[root@localhost php-5.3.28]# make
[root@localhost php-5.3.28]# make install
[root@localhost php-5.3.28]# cd /usr/local/php/etc/
[root@localhost etc]# cp php-fpm.conf.default php-fpm.conf
[root@localhost etc]# vim php-fpm.conf
……                            // 省略部分信息
pid = run/php-fpm.pid
user = php
group = php
```

```
listen = 0.0.0.0:9000
pm.max_children = 50
pm.start_servers = 20
pm.min_spare_servers = 5
pm.max_spare_servers = 35
[root@localhost etc]# /usr/local/php/sbin/php-fpm
[root@localhost etc]# netstat -tunlp | grep 9000
tcp   0  0 0.0.0.0:9000   0.0.0.0:*        LISTEN      28383/php-fpm
[root@localhost etc]# mkdir –p /var/www/html/webphp
[root@localhost etc]# vim /var/www/html/webphp/index.php
<?
phpinfo();
?>
```

（4）访问 Nginx 测试。

通过浏览器访问 192.168.0.10/index.php，能够看到 PHP 的界面。

3. 部署 MySQL 主从数据库

部署过程略。

按照 PHP 程序员的要求在 MySQL 主服务器上创建 friends 数据库，并使创建 PHP 程序的用户对 friends 数据库具有读写权限。

```
[root@localhost ~]# mysql -u root -p
Enter password:
……                       // 省略部分输出
mysql> create database friends;
Query OK, 1 row affected (0.00 sec)

mysql> grant all privileges on friends.* to 'php'@'192.168.0.%' identified by 'yourpasswd';
    Query OK, 0 rows affected (0.01 sec)
```

4. MFS 部署并挂载

部署过程略。

将 MFS 文件系统挂载在前端 PHP 服务器的 /var/www/html/webphp/uploads/photos 目录下。

5. 发布上线

（1）通知上级与部署人员，可以开始上线发布。

通过 SVN 服务器的 sysadmin 账户部署代码至 Nginx 服务器及 PHP 服务器，注意 Nginx 和 PHP 服务器配置的服务目录分别是 /usr/local/nginx/html/webphp 和 /var/www/html/webphp。

```
[root@localhost ~]# cd /usr/local/nginx/html
[root@localhost html]# svn co svn://192.168.0.172/webphp
// 根据提示使用账户 sysadmin 登录，即可部署代码
```

```
[root@localhost html]# ls
webphp
```

PHP 服务器的部署方式与 Nginx 服务器类似。

（2）通知上级与测试部门上线完毕，可以开始测试。

5.3 案例扩展

公司社交网站上线后，上级领导对网站的运维要求进一步提高，希望公司网站的可靠性与性能一并提高，具体要求如下：

（1）使用 Keepalived 对 Nginx 服务器做热备，原来的服务器真实 IP 地址修改为 192.168.0.9，VIP 地址采用 192.168.0.10，另外增加一台 IP 地址为 192.168.0.11 的 Nginx 服务器作为热备。

（2）增加一台 PHP 服务器，前端 Nginx 服务器通过 upstream 模块实现对 PHP 服务器的负载均衡。

（3）对 Nginx 服务器与 PHP 服务器等进行性能调优，对比集群扩展前后、优化前后的压力测试数据。

本章总结

- 公司的社交网站采用 PHP 开发，搭建 SVN 服务器进行版本控制，社交网站的第一个版本部署在 LNMP 平台之上。
- 社交网站项目包含用户的相册功能，允许用户上传照片，上传照片使用 MFS 分布式文件系统来实现，将 MFS 挂载在 PHP 服务器的相关目录下。
- 使用 keepalived 对 Nginx 服务器做热备，增加一台 PHP 服务器，前端 Nginx 服务器通过 upstream 模块实现对 PHP 服务器的负载均衡。
- 对 Nginx 服务器与 PHP 服务器等进行性能调优，对比集群扩展前后、优化前后的压力测试数据。

随手笔记

第 6 章

大型网站架构

技能目标

- 理解 PV 的概念
- 实现百万 PV 网站架构
- 理解千万 PV 网站架构
- 会配置 Redis 主从复制
- 会配置 RabbitMQ 集群

本章导读

网站架构一般认为是根据客户需求分析的结果,准确定位网站目标群体,设计网站的整体架构,规划、设计网站栏目及其内容,制定网站开发流程的顺序,最大限度地进行高效资源分配与管理的设计。本章将介绍百万 PV 网站和千万 PV 网站架构的案例。

知识服务

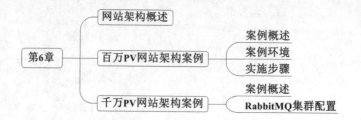

6.1 网站架构概述

网站架构一般认为是根据客户需求分析的结果，准确定位网站目标群体，设定网站的整体架构，规划、设计网站栏目及其内容，制定网站开发流程的顺序，最大限度地进行高效资源分配与管理的设计。

网站架构分很多种，一般我们平常所说的是软件方面的架构，例如：

- 前端使用了什么代理服务器？
- Web 服务器用的是什么？
- 中间又使用了什么缓存服务器？
- 数据库服务器用的是什么？
- 代码又是基于什么框架开发的？
- 这个网站架构每天访问量有多少（通常所说的 PV）？并发数为多少？

PV（Page View，页面浏览量）即点击量，通常是衡量一个网络新闻频道或网站甚至一条网络新闻的主要指标。PV 之于网站，就像收视率之于电视，从某种程度上已成为投资者衡量商业网站表现的最重要的尺度。

对 PV 的解释是，一个访问者在 24 小时（0 点到 24 点）内到底看了网站的几个页面。这里需要注意的是，同一个人浏览网站的同一个页面，不重复计算 PV 量，点 100 次也只算 1 次。

如何计算 PV 呢？当一个访问者访问的时候，记录他所访问的页面和对应的 IP，然后确定这个 IP 今天访问了这个页面没有。如果网站到了 23 点，IP 有 10 万条的话，每个访问者平均访问了 3 个页面，那么 PV 表的记录就要有 30 万条。

查询网站 PV 可以访问 Alexa.cn，免费提供网站访问量查询、网站浏览量查询和排名变化趋势数据查询。

6.2 百万 PV 网站架构案例

1．案例概述

本案例设计采用四层模式实现，主要分为前端反向代理层、Web 层、数据库缓存层和数据库层。前端反向代理层采用主备模式，Web 层采用集群模式，数据库缓存层采用主备模式，数据库层采用主从模式。

为了更接近生产环境，我们采用两台实体机部署此次环境，将前端反向代理层、数据库缓存层、数据库层部署在实体机上，只将 Web 层部署在 kvm 虚拟机当中。同时将每一层都做了高可用架构，以保证业务的稳定性。

拓扑架构如图 6.1 所示，实线是正常情况下的数据流向连接，虚线是异常情况下的数据流向连接。

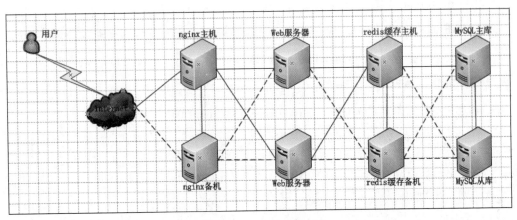

图 6.1　百万 PV 网站架构

2．案例环境

案例环境如表 6-1 所示。

表 6-1　案例环境

主机名	IP 地址	系统版本	用途
master	外网：223.202.18.81 内网：192.168.10.81	CentOS 6.5（64 位）	前端反向代理主机、redis 缓存主机、MySQL 数据主库
backup	外网：223.202.18.82 内网：192.168.10.82	CentOS 6.5（64 位）	前端反向代理备机、redis 缓存备机、MySQL 数据从库
tomcat-node1	内网：192.168.10.83	CentOS 6.5（64 位）	Web 服务
tomcat-node2	内网：192.168.10.84	CentOS 6.5（64 位）	Web 服务

因为需要安装一些软件，所以上述两台反向代理服务器要求能连接外网。这里前端使用 Keepalived 作为高可用软件，虚拟 IP 设置为 223.202.18.80 和 192.168.10.80，这里设置一个虚拟的内网 IP 是为了内部各个系统之间进行通信。

由于是搭建学习测试环境，所以关闭了上述所有服务器的防火墙和 Selinux。如果是生产环境建议开启防火墙。所有服务都是默认配置的，没有进行任何优化，需要撑起更大的访问量必须要进行优化。

另外，生产环境建议将 MySQL 的分区设置为 xfs 类型，因为在大多数场景下，整体 IOPS 表现还是要比 ext4 来的更高、更稳定，延迟也更小。

3. 实施步骤

（1）前端两台反向代理服务器安装 epel 和 nginx 源。

```
rpm -ivh http://mirrors.ustc.edu.cn/fedora/epel/6/x86_64/epel-release-6-8.noarch.rpm
```

修改 /etc/yum.repos.d/epel.repo 文件，在 [epel] 字段里面增加如下内容：

```
baseurl=http://mirrors.ustc.edu.cn/fedora/epel/6/$basearch
```

注释 mirrorlist 这行

```
rpm –ivh http://nginx.org/packages/centos/6/noarch/RPMS/nginx-release-centos-6-0.el6.ngx.noarch.rpm
```

执行如下命令安装 keepalived 和 nginx 软件：

```
yum install keepalived nginx
```

前端反向代理主机的 /etc/keepalived/keepalived.conf 配置文件如下：

```
! Configuration File for keepalived

global_defs {
  router_id NGINX_HA
}

vrrp_script nginx {
  script "/opt/shell/nginx.sh"
  interval 2
}

vrrp_instance VI_1 {
  state MASTER
  interface br0
  virtual_router_id 51
  priority 100
  advert_int 1
  authentication {
    auth_type PASS
    auth_pass 1111
  }

  track_script {
    nginx
  }

  virtual_ipaddress {
    223.202.18.80
    192.168.10.80
  }
}
```

在配置文件里面载入了一个 nginx.sh 脚本，脚本的含义是如果 keepalived 已启动，那么每两秒检查并启动一次 nginx 服务。

创建 /opt/shell/nginx.sh 文件，内容如下：

```
#!/bin/bash
k='ps -ef | grep keepalived | grep -v grep | wc -l'
if [ $k -gt 0 ];then
/etc/init.d/nginx start
else
/etc/init.d/nginx stop
fi
```

脚本文件增加可执行权限：

chmod +x /opt/shell/nginx.sh

启动 keepalived 服务：

/etc/init.d/keepalived start

查看 /var/log/message 日志信息是否有配置文件里面设定的虚拟 IP 绑定，或者使用 ip a 命令查看，确定 nginx 服务也是正常运行的。

下面开始前端反向代理备机的 keepalived 的配置，内容如下：

```
! Configuration File for keepalived

global_defs {
    router_id NGINX_HA
}

vrrp_script nginx {
    script "/opt/shell/nginx.sh"
    interval 2
}

vrrp_instance VI_1 {
    state BACKUP        // 修改
    interface br0
    virtual_router_id 51
    priority 99         // 修改
    advert_int 1
    authentication {
        auth_type PASS
        auth_pass 1111
    }
    track_script {
        nginx
```

```
        }
    virtual_ipaddress {
       223.202.18.80
       192.168.10.80
        }
    }
```

其实可以完全拷贝主机上的，然后略做修改。脚本文件也可以完全拷贝过来，不用修改。

备机启动 keepalived 服务后，nginx 服务也会启动。

关闭主机的 keepalived 服务，查看虚拟 IP 是否可以自动漂移到备机上？确认 keepalived 服务配置没问题后，可以再次将主机的 keepalived 服务开启，虚拟 IP 又会漂移到主机上。

（2）两台前端反向代理主机和备机开始配置 nginx 服务，主机和备机保持一模一样的配置就可以。

打开 /etc/nginx/nginx.conf 主配置文件，增加如下内容，在 include 行上面。

```
upstream tomcat_pool {
    server 192.168.10.83:8080;
    server 192.168.10.84:8080;
}

server {
    listen 80;
    server_name 223.202.18.80;
    location / {
    root html;
    index index.html index.htm;
    proxy_pass    http://tomcat_pool;
    proxy_set_header X-Real-IP $remote_addr;
    }
}
```

在重新加载前，测试 nginx 配置文件是否有语法问题。

在浏览器中输入 http://223.202.18.80 测试转发是否正常，如果出现 502，就说明转发正常。

（3）开始安装 Web 服务器，直接在一台 Web 服务器上执行。待测试无问题后再将整个 tomcat 目录拷贝过去。

上传并安装 jdk 过程略（两台 Web 服务器都执行），上传 tomcat 软件包过程略。

解压 tomcat 软件包：

```
tar zxvf apache-tomcat-7.0.52.tar.gz
mv  apache-tomcat-7.0.52 /usr/local/apache
```

在 tomcat-node1 的 /etc/hosts 文件中增加如下内容：

192.168.10.83 tomcat-node1

在 tomcat-node2 的 /etc/hosts 文件中增加如下内容：

192.168.10.84 tomcat-node2

（4）安装 mysql 数据库并导入数据，这里我们使用 yum 方式安装 5.5 版本。两台数据库上都要安装，但只在主库上导入数据。

首先安装 remi 源：

rpm -ivh http://rpms.famillecollet.com/enterprise/remi-release-6.rpm

将 /etc/yum.repos.d/remi.repo 文件中的 [remi] 字段里的 baseurl 行注释去掉，对 mirrorlist 行注释，最后将 enabled=0 修改为 1。

安装 MySQL 数据库：

yum install mysql mysql-devel mysql-server

启动 MySQL 数据库：

/etc/init.d/mysqld start

开始创建库导入数据：

create database slsaledb
mysql -u root slsaledb < slsaledb-2014-4-10.sql

上述导入命令没有密码，但是生产环境中 MySQL 必须设置 root 密码。

授权 Web 应用连接数据库的 IP 地址、用户名和密码，生产环境不建议使用 root 用户。

grant all on slsaledb.* to root@'192.168.10.%' identified by '123456';

MySQL 主从稍后配置。

（5）开始配置 Web 服务并测试。

将项目上传到 tomcat 的 webapps 目录下，工程名称为 SLSaleSystem，修改连接数据库的配置文件 jdbc.properties，如图 6.2 所示。

```
[root@tomcat-node1 ~]# more /usr/local/tomcat/webapps/SLSaleSystem/WEB-INF/classes/jdbc.properties
driverClassName=com.mysql.jdbc.Driver
url=jdbc\:mysql\://192.168.10.80\:3306/slsaledb?useUnicode\=true&characterEncoding\=UTF-8
uname=root
password=123456
minIdle=10
maxIdle=50
initialSize=5
maxActive=100
maxWait=100
removeAbandonedTimeout=180
removeAbandoned=true
```

图 6.2　修改连接数据库的配置文件

将连接数据库的 IP 地址、用户名、密码修改如上图所示。

修改 tomcat-node1 的主配置文件 /usr/local/tomcat/conf/server.xml，在 </Host> 字段上增加如下内容，以便用户访问的时候不用输入项目名称。

```
<Context path="" docBase="SLSaleSystem" reloadable="true" debug="0">
</Context>
```

启动 tomcat 服务：

```
/usr/local/tomcat/bin/startup.sh
```

在浏览器上测试 tomcat 节点是否能正常访问？注意目前测试还不是通过前端的 nginx 去访问的。如图 6.3 所示，测试登录，输入用户名 admin，密码 123456 进行测试。

图 6.3　测试登录

因为 Web 应用使用的是集群，所以将配置无问题的 tomcat 目录拷贝到另外一台 Web 服务器上。不用作任何修改，直接启动即可。

接下来配置两台 Web 服务器的会话复制。因为一个登录用户的会话在某台 Web 服务器上，如果此时这台 Web 服务器故障，那么用户的会话就会丢失，用户在点击的时候则需要重新再登录一次，影响使用。这里因为是百万的 PV 架构，规模还不是很大，所以用 tomcat 自身的方案即可解决。

打开第一台的 tomcat 主配置文件 /usr/local/tomcat/conf/server.xml，将 <Engine name="Catalina" defaultHost="localhost"> 修改为：

```
<Engine name="Catalina" defaultHost="localhost" jvmRoute="node1">
```

同样地，第二台的 tomcat 主配置文件修改为：

```
<Engine name="Catalina" defaultHost="localhost" jvmRoute="node2">
```

将两台 Web 服务器的 /usr/local/tomcat/conf/server.xml 中下面这行的注释去掉。

```
<Cluster className="org.apache.catalina.ha.tcp.SimpleTcpCluster"/>
```

最后在两台 Web 服务器项目中的 /usr/local/tomcat/webapps/SLSaleSystem/WEB-INF/web.xml 文件的 </web-app> 字段上面增加如下内容：

```
<distributable/>
```

为了测试方便，编写一个测试页面。路径如下：

/usr/local/tomcat/webapps/SLSaleSystem/test.jsp

第一台 tomcat 内容如下：

```
SessionID:<%=session.getId()%>
<BR>
SessionPort:<%=request.getServerPort()%>
<%
out.println("This is Tomcat Server 192.168.10.83!");
%>
```

第二台 tomcat 内容如下：

```
SessionID:<%=session.getId()%>
<BR>
SessionPort:<%=request.getServerPort()%>
<%
out.println("This is Tomcat Server 192.168.10.84!");
%>
```

最后通过前端反向代理虚拟 IP 地址访问测试 http://223.202.18.80，如图 6.4 所示。

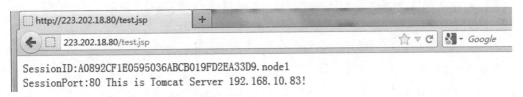

图 6.4　Session 访问测试（1）

因为 Nginx 默认采用的是轮询的访问方式，如果后续业务需要也可以设置为 ip_hash 模式。手动刷新网页，如图 6.5 所示。

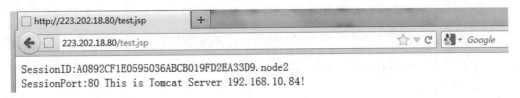

图 6.5　Session 访问测试（2）

从上图中我们看到 SessionID 的值没有变化，但是后端响应的服务器已经改变，说明 Web 服务器集群配置完毕。

（6）接下来我们开始安装并配置 redis 主从缓存服务器。

1）Redis 简介

Redis 是一个高性能的 key-value 数据库，和 Memcached 类似，但它支持存储的 value 类型更多。与 Memcached 一样，为了保证效率，数据都是缓存在内存中的。

区别是 redis 会周期性地把更新的数据写入磁盘或者把修改操作写入追加的记录文件中，并且在此基础上实现了 master-slave（主从）同步。

Redis 的出现，很大程度上补偿了 Memcached 这类 key-value 存储的不足，在部分场合可以对关系数据库起到很好的补充作用。

关于 Redis 的详细介绍请上课工场 APP 或官网 kgc.cn 观看视频。

2）安装并配置 redis 主从

两台服务器上都执行安装命令：

```
yum install redis
```

修改主缓存服务器的 redis 主配置文件 /etc/redis.conf 中的 redis 监听端口，因为默认是 127.0.0.1，修改为 bind 192.168.10.81。

主缓存服务器启动服务：

```
/etc/init.d/redis start
```

客户端连接主缓存服务器：

```
redis-cli -h 192.168.10.81 6379
```

手动插入一条数据并且查询：

```
redis 192.168.10.81:6379> set name test
OK
redis 192.168.10.81:6379> get name
"test"
```

从缓存服务器修改主配置文件 /etc/redis.conf 中的 redis 监听端口，修改为 bind 192.168.10.82。

然后在 # slaveof <masterip><masterport> 行下面增加如下内容：

```
slaveof 192.168.10.81 6379
```

启动从缓存服务器的 redis 服务：

```
/etc/init.d/redis start
```

在从缓存服务器上，使用 redis 客户端连接主缓存 redis 服务，查询刚在主缓存服务上插入的 name 值。

```
[root@nignx-backup ~]# redis-cli –h 192.168.10.81 -p 6379
redis 192.168.10.81:6379> get name
"test"
```

结果说明主从缓存服务器同步正常。测试没问题后删除刚刚插入的值，使用命令 del name 即可。

当然这只是 redis 最简单的配置，一般情况下运维工程师只需要搭建好服务即可，至于 Web 服务器里面的代码如何去连接 redis 是开发人员的工作。

下面是一段代码示例，仅供参考：

```xml
<bean id="jedisPoolConfig" class="redis.clients.jedis.JedisPoolConfig">
    <property name="maxActive" value="90" />
    <property name="maxIdle" value="5" />
    <property name="maxWait" value="1000" />
    <property name="testOnBorrow" value="true" />
</bean>
<bean id="jedisPool" class="redis.clients.jedis.JedisPool" destroy-method="destroy" >
    <constructor-arg ref="jedisPoolConfig"/>
    <constructor-arg value=" 缓存服务器地址 "/>
    <constructor-arg value=" 缓存服务器端口 "/>
</bean>
<bean id="redisAPI" class="org.slsale.common.RedisAPI">
    <property name="jedisPool" ref="jedisPool"/>
</bean>
```

将主缓存服务器的 redis 服务重启，那么之前的所有缓存都会丢失。

当开发人员在代码里面写了连接 redis 缓存服务器时，运维人员需要关注的是使用缓存是否生效。

（7）配置 MySQL 主从配置。

下面是配置主从同步过程，同步的库名为 slsaledb。

首选主库需要开启 binlog 日志，在打开主库的 /etc/my.cnf 文件里面增加如下内容：

```
server-id=1
log-bin=binlog
binlog-do-db=slsaledb
```

重新启动主库：

```
/etc/init.d/mysqld restart
```

登录主库并授权，允许从库的指定用户从主库读取日志：

```
mysql> grant replication slave on *.* to 'rep'@'192.168.10.82' identified by '123456';
```

将主库上 slsaledb 库导出为 sql 文件。

```
mysqldump –u root slsaledb > slsaledb.sql
```

将 slsaledb.sql 文件拷贝到从库上面进行导入，导入之前先在从库上创建 slsaledb 库。

```
create database slsaledb;
mysql –u root < slsaledb.sql
```

在主库上执行如下命令，记录二进制文件和 position 信息，因为从库同步的时候需要。

```
mysql> show master status;
+---------------+----------+--------------+------------------+
| File          | Position | Binlog_Do_DB | Binlog_Ignore_DB |
+---------------+----------+--------------+------------------+
| binlog.000001 |      571 | slsaledb     |                  |
+---------------+----------+--------------+------------------+
1 row in set (0.00 sec)
```

修改从库的 /etc/my.cnf 文件，增加如下内容：

server-id=2

重新启动从库：

/etc/init.d/mysqld restart

登录从库并且开始同步：

```
mysql> stop slave
mysql> change master to master_host='192.168.15.81',master_user='rep',master_password=
    '123456',master_log_file='binlog.000001',master_log_pos=571;
```

查看从库状态，下面加粗字体必须是 Yes 状态。

```
mysql> show slave status\G;
*************************** 1. row ***************************
             Slave_IO_State: Waiting for master to send event
                Master_Host: 192.168.15.81
                Master_User: rep
                Master_Port: 3306
              Connect_Retry: 60
            Master_Log_File: binlog.000001
        Read_Master_Log_Pos: 571
             Relay_Log_File: localhost-relay-bin.000003
              Relay_Log_Pos: 606
      Relay_Master_Log_File: binlog.000001
           Slave_IO_Running: Yes            （代表网络正常）
          Slave_SQL_Running: Yes            （代表表结构正常）
            Replicate_Do_DB:
        Replicate_Ignore_DB:
         Replicate_Do_Table:
     Replicate_Ignore_Table:
    Replicate_Wild_Do_Table:
Replicate_Wild_Ignore_Table:
                 Last_Errno: 0
                 Last_Error:
               Skip_Counter: 0
```

```
                  Exec_Master_Log_Pos: 292
                    Relay_Log_Space: 327
                    Until_Condition: None
                     Until_Log_File:
                      Until_Log_Pos: 0
                 Master_SSL_Allowed: No
                 Master_SSL_CA_File:
                 Master_SSL_CA_Path:
                    Master_SSL_Cert:
                  Master_SSL_Cipher:
                     Master_SSL_Key:
              Seconds_Behind_Master: 0
1 row in set (0.00 sec)
ERROR:
No query specified
```

如果不放心主从同步是否正常，则使用手动测试也可以。

在主库 slsaledb 上新建一个表并插入一些记录，确认从库是否同步了这张表过去，数据是否正常。

本案例数据库的架构简单了一点，配置文件也是默认的没有经过优化，如果数据库的压力很大，可以考虑优化，也可以使用读写分离架构，或者采用多主多从模式。这些在后面的千万 PV 架构案例中会体现出来。

6.3　千万 PV 网站架构案例

6.3.1　案例概述

本案例来源于国内某大型网站，架构图如图 6.6 所示。

将图 6.6 划分为几个区域，每个区域都组成了一个集群，且都是由一台服务器扩展到了多台服务器。详细的介绍请上课工场 APP 或官网 kgc.cn 观看视频。

这个架构也比较庞大，一个网站每天能支撑多少的 PV 访问量和它的架构是密不可分的，但要注意并非一味拼命地堆机器，那样只会增加运营成本和管理难度。正确的做法是去优化每台服务器的性能、服务或者接口，来提升资源最大的利用率。当优化后到达一定的峰值时，再去考虑横向扩展性。想想 QQ 早期的时候只有一台服务器，但是他们每个星期都要优化一次程序。QQ 的在线人数从 1000 人发展到 2000 人，最后再到 4000 人，人数发生了翻天覆地的变化，他们并没有盲目地去买服务器，而是不断地优化代码和服务。

在现实中我们无法重现大规模用户并发的场景，只能通过网站压力软件来模拟，一般公司都会有专业的测试团队来测试网站上线前的功能性和健壮性。JMeter 这个工具不需要运维人员很熟悉，但是至少要知道如何使用。

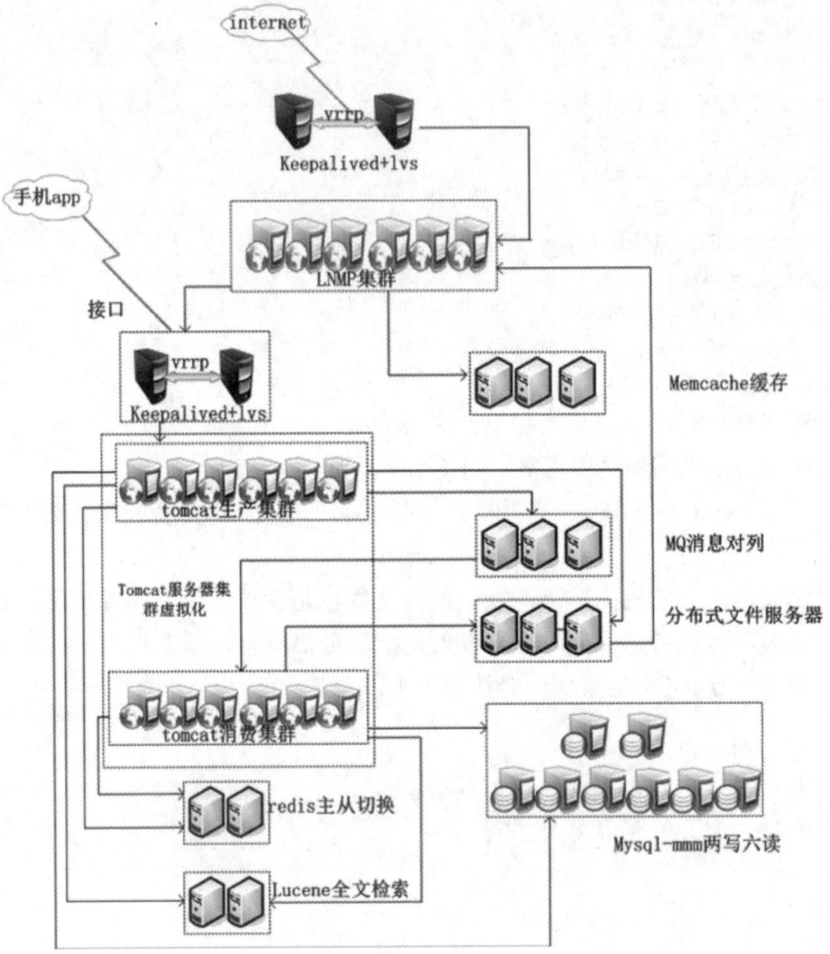

图 6.6 千万 PV 网站架构

6.3.2 RabbitMQ 集群配置

1. RabbitMQ 是什么？

MQ（Message Queue，消息队列）是一种应用程序对应用程序的通信方法。应用程序通过读写出入队列的消息（针对应用程序的数据）来通信，而无需专用链接来连接它们。消息传递指的是程序之间通过在消息中发送数据进行通信，而不是通过直接调用彼此来通信，直接调用通常是用于诸如远程过程调用的技术。排队指的是应用程序通过队列来通信。队列的使用除去了接收和发送应用程序同时执行的要求。

RabbitMQ 是目前流行的开源消息队列系统，用 Erlang 语言开发。RabbitMQ 是 AMQP（高级消息队列协议）的标准实现。如果不熟悉 AMQP，直接看 RabbitMQ 的文档会比较困难。不过它也只有几个关键概念，这里简单介绍如下：

Broker：简单来说就是消息队列服务器实体。

Exchange：消息交换机，它指定消息按什么规则，路由到哪个队列。

Queue：消息队列载体，每个消息都会被投入到一个或多个队列中。

Binding：绑定，它的作用就是把 Exchange 和 Queue 按照路由规则绑定起来。

Routing Key：路由关键字，Exchange 根据这个关键字进行消息投递。

Vhost：虚拟主机，一个 Broker 里可以开设多个 Vhost，用作不同用户的权限分离。

Producer：消息生产者，就是投递消息的程序。

Consumer：消息消费者，就是接受消息的程序。

Channel：消息通道，在客户端的每个连接里，可建立多个 Channel，每个 Channel 代表一个会话任务。

消息队列的使用过程如下：

（1）客户端连接到消息队列服务器，打开一个 Channel。

（2）客户端声明一个 Exchange，并设置相关属性。

（3）客户端声明一个 Queue，并设置相关属性。

（4）客户端使用 Routing Key，在 Exchange 和 Queue 之间建立好绑定关系。

（5）客户端投递消息到 Exchange。

Exchange 接收到消息后，就根据消息的 Key 和已经设置的 Binding，进行消息路由，将消息投递到一个或多个队列里。

Exchange 也有几个类型，完全根据 Key 进行投递的叫做 Direct 交换机，例如，绑定时设置了 Routing Key 为 abc，那么客户端提交的消息，只有设置了 Key 为 abc 的才会投递到队列。对 Key 进行模式匹配后进行投递的叫做 Topic 交换机，符号"#"匹配一个或多个词，符号"*"匹配一个词。例如 abc.# 匹配 abc.def.ghi，abc.* 只匹配 abc.def。还有一种不需要 Key 的，叫做 Fanout 交换机，它采取广播模式，当一个消息进来时，投递到与该交换机绑定的所有队列。

2. RabbitMQ 使用场景

在项目中，将一些无需即时返回且耗时的操作提取出来，进行异步处理，而这种异步处理的方式大大地节省了服务器的请求响应时间，从而提高了系统的吞吐量。

RabbitMQ 支持消息的持久化，也就是数据写在磁盘上。为了数据安全考虑，大多数企业都会选择持久化。当然如果觉得不需要消息持久化，那么使用内存节点即可！其实最合适的方案就是既有内存节点，又有磁盘节点，下面的案例就是这样一个例子。消息队列持久化包括 3 个部分：

（1）Exchange 持久化，在声明时指定 durable => 1。

（2）Queue 持久化，在声明时指定 durable => 1。

（3）消息持久化，在投递时指定 delivery_mode => 2（1 是非持久化）。

如果 Exchange 和 Queue 都是持久化的，那么它们之间的 Binding 也是持久化的。

如果 Exchange 和 Queue 两者之间只有一个持久化，而另一个非持久化，就不允许建立绑定。

RabbitMQ 的结构图如图 6.7 所示。

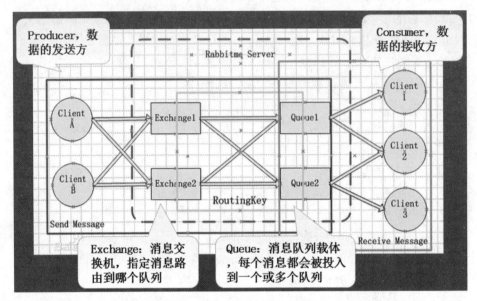

图 6.7　RabbitMQ 的结构图

3. RabbitMQ 实际应用

RabbitMQ 是用 Erlang 开发的，集群非常方便，因为 Erlang 天生就是一门分布式语言，但其本身并不支持负载均衡。

RabbitMQ 模式大概分为以下三种：

（1）单一模式。

（2）普通模式（默认的集群模式）。

（3）镜像模式（把需要的队列做成镜像队列，存在于多个节点，属于 RabbitMQ 的 HA 方案，在对业务可靠性要求较高的场合中比较适用）。

要实现镜像模式，需要先搭建一个普通集群模式，在这个模式的基础上再配置镜像模式以实现高可用。

RabbitMQ 的集群节点包括内存节点、磁盘节点。RabbitMQ 支持消息的持久化，也就是数据写在磁盘上，最合适的方案就是既有内存节点，又有磁盘节点。

4. 实现步骤

设计架构模式：在一个集群里，有三台服务器，其中一台使用磁盘模式，另两台使用内存模式。两台内存模式的节点无疑速度更快，因此通过客户端连接访问它们。但是客户端不可能分别连接两个内存节点，肯定是通过前端反向代理去轮询分发请求。如果担心前端反向代理服务器故障，可以通过 Keepalived 软件做一个高可用架构。而磁盘模式的节点，由于磁盘 IO 相对较慢，因此仅作数据备份使用。

注意这里请将三台服务器都连接上互联网并安装软件包。另外 RabbitMQ 集群节点必须在同一个网段里，如果是跨广域网，效果就会变差。

RabbitMQ 集群具体配置信息如表 6-2 所示。

表 6-2　RabbitMQ 集群具体配置

IP 地址	主机名	操作系统	防火墙和 SElinux	用途
192.168.2.19	rabbitmq01	CentOS6.5（64 位）	关闭	磁盘节点
192.168.2.20	rabbitmq02	CentOS6.5（64 位）	关闭	内存节点
192.168.2.23	rabbitmq03	CentOS6.5（64 位）	关闭	内存节点

（1）首先我们需要配置三个节点的 hosts 文件，将如下内容分别加入到三台服务器上。

```
192.168.2.19  rabbitmq01
192.168.2.20  rabbitmq02
192.168.2.23  rabbitmq03
```

（2）三个节点配置 yum 软件源，安装 rabbitmq 软件。

rpm -ivh http://mirrors.ustc.edu.cn/fedora/epel/6/x86_64/epel-release-6-8.noarch.rpm

修改 /etc/yum.repos.d/epel.repo 文件，在 [epel] 字段里面增加如下内容：

baseurl=http://mirrors.ustc.edu.cn/fedora/epel/6/$basearch

注释 mirrorlist 这行，并安装 rabbitmq-server 软件：

yum install rabbitmq-server –y

（3）分别查看三个节点并添加管理服务，最后启动 rabbitmq 服务：

```
/usr/lib/rabbitmq/bin/rabbitmq-plugins list                        // 查看有哪些服务
/usr/lib/rabbitmq/bin/rabbitmq-plugins enable rabbitmq_management  // 启动管理服务
/etc/init.d/rabbitmq-server start                                  // 启动 rabbitmq 服务
```

使用 netstat -an | grep 5672 进行查看，有如下三个端口开放说明正常。其中 15672 和 55672 都是 rabbitmq 的管理端口，5672 则是和生产者、消费者通信的端口。

Rabbitmq 的集群是依赖于 Erlang 的集群来工作的，所以必须先构建起 Erlang 的集群环境。在 Erlang 的集群中，各节点是通过一个 magic cookie 来实现的，这个 cookie 存放在 /var/lib/rabbitmq/.erlang.cookie 中，文件是 400 的权限。所以必须保证各节点 cookie 保持一致，否则节点之间就无法通信。

将其中一台节点上的 .erlang.cookie 值复制下来保存到其他节点上，或者使用 scp 的方法也可以，但是要注意文件的权限和属主属组。

现在三个节点的 rabbitmq 服务都是开启的，但是每个节点的 .erlang.cookie 文件中内容的值都不一样，因此我们需要将两个内存节点的 rabbitmq 服务停止。

然后将磁盘节点上的 /var/lib/rabbitmq/.erlang.cookie 文件中的内容复制到两个内存节点上的 .erlang.cookie 中。再次启动两个内存节点的 rabbitmq 服务，并将它们加入一个集群当中。

将 rabbitmq02、rabbitmq03 作为内存节点与 rabbitmq01 磁盘节点连接起来，在 rabbitmq02 上执行如下命令：

```
[root@rabbitmq02 ~] # rabbitmqctl stop_app
Stopping node rabbit@rabbitmq02 ...
...done.
[root@rabbitmq02 ~] # rabbitmqctl join_cluster --ram rabbit@rabbitmq01
Clustering node rabbit@rabbitmq02 with rabbit@rabbitmq01 ...
...done.
[root@rabbitmq02 ~] # rabbitmqctl start_app
Starting node rabbit@rabbitmq02 ...
...done.
```

在 rabbitmq03 上也同样执行上述命令。

上述命令会先停掉 rabbitmq 应用，然后调用 cluster 命令，将 rabbitmq02 连接到 rabbitmq01，使两者成为一个集群，最后启动 rabbitmq 应用。

在这个 cluster 命令下，rabbitmq02 和 rabbitmq03 是内存节点，rabbitmq01 是磁盘节点。而且默认 rabbitmq 启动后是磁盘节点。如果要使 rabbitmq02 和 rabbitmq03 都是磁盘节点，去掉 --ram 参数即可。只要在节点列表里包含了自己，它就成为了一个磁盘节点。在 RabbitMQ 集群里，至少有一个磁盘节点存在。

上面已完成了配置 RabbitMQ 默认集群模式，但并不保证队列的高可用性，尽管交换机、绑定这些可以复制到集群里的任何一个节点，但是队列内容不会被复制，虽然该模式解决了一部分节点压力，但队列节点宕机会直接导致该队列无法使用，只能等待重启，所以要想在队列节点宕机或故障也能正常使用，就要复制队列内容到集群里的每个节点，这就需要创建镜像队列。

下面来看如何采用镜像模式来解决复制的问题，从而提高可用性。

使用 rabbitmq 镜像功能，需要基于 rabbitmq 策略来实现。我们先在集群中的一台内存节点上创建一个策略。

打开浏览器输入 http://192.168.2.19:55672，它会自动将 55672 端口改变为 15672 端口。

输入默认的 Username：guest，输入默认的 Password：guest。登录后会出现如图 6.8 所示的界面。

从图中我们看到了很多信息，比如最后面的 Type 类型表示哪些节点是基于磁盘和内存存储的。

点击最上面的菜单栏 Admin 按钮，如图 6.9 所示。

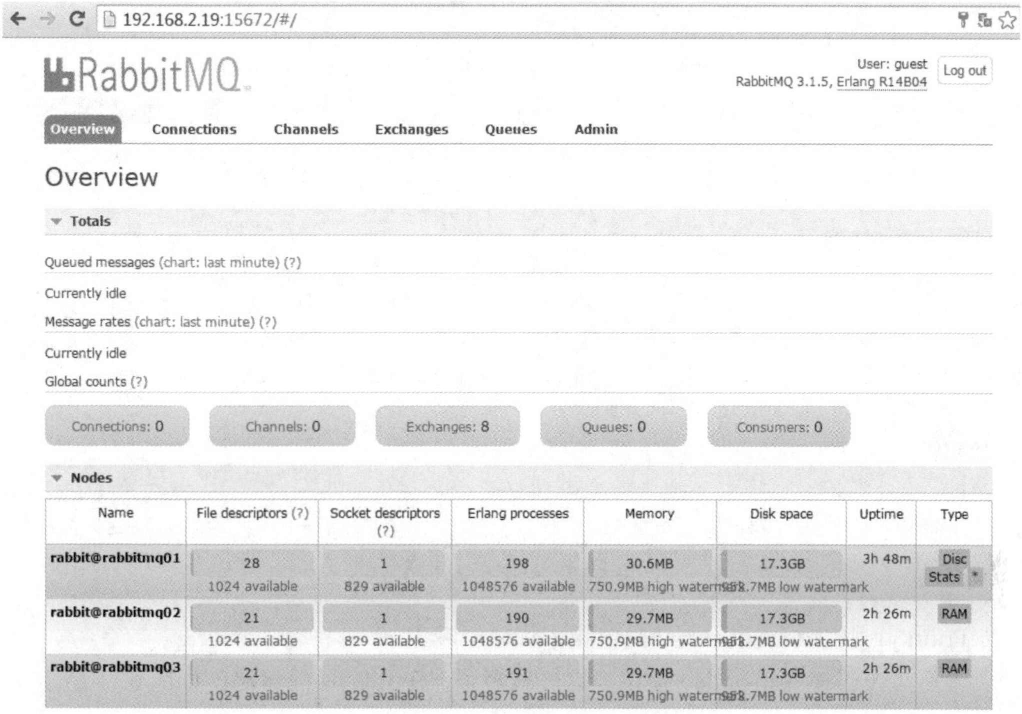

图 6.8　登录后的界面

图 6.9　菜单栏 Admin

点击右边的 Policies 按钮，然后创建一个策略，分别写入 Name、Pattern、Definition 对应的值，如图 6.10 所示。

Name：策略名称。

Pattern：匹配的规则，这里表示匹配 a 开头的队列，如果是匹配所有的队列，那就是 ^。

Definition：使用 ha-mode 模式中的 all，也就是同步所有匹配的队列。还有个问号，我们也可以点击查看，类似于帮助文档。

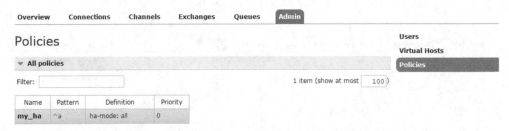

图 6.10　创建策略（1）

Priority：优先级，默认是 0，值越大优先级越大。

最后点击添加 Add policy 按钮，这样就多了一个 my_ha 的策略，如图 6.11 所示。

图 6.11　创建策略（2）

此时分别登录 rabbitmq02、rabbitmq03 两个节点的控制台，也可以看到上面添加的这个策略，如图 6.12 和图 6.13 所示。

图 6.12　查看策略（1）

图 6.13　查看策略（2）

策略已经添加了，然后就开始添加队列，最后还需要创建消息。

下面在任意一台节点上添加一个 Queues 队列，点击 Queues 按钮，输入 Name 和 Arguments 参数的值，别的值按其默认值即可，如图 6.14 所示。

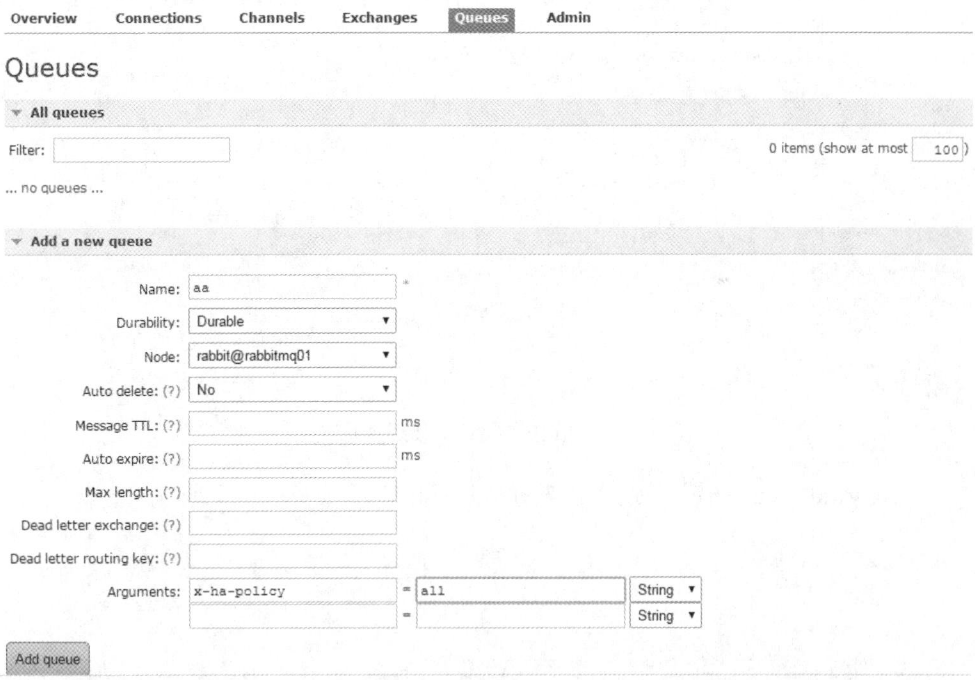

图 6.14　添加队列（1）

Name：队列的名称。

Durability：队列是否持久化（Durable 是持久化）。

Node：消息队列的节点。

Auto delete：自动删除。

Arguments：使用的策略类型。

点击 Add queue 按钮，会出现一个名为 aa 的队列，将鼠标指向 +2 可以显示出另外两台消息节点，如图 6.15 所示。

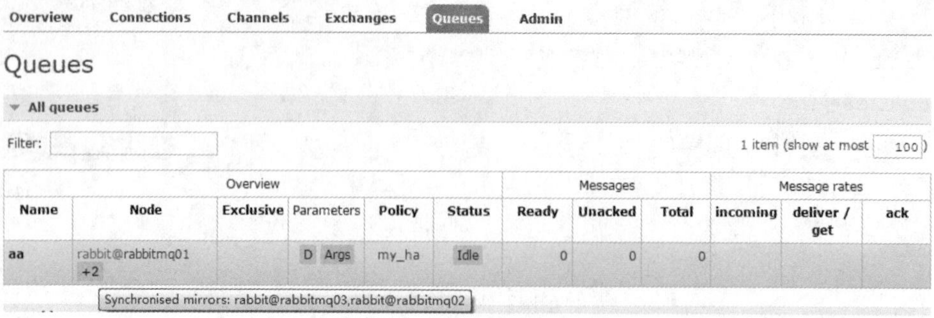

图 6.15　添加队列（2）

接下来再创建一个队列，不填写 Arguments 的值，如图 6.16 所示。

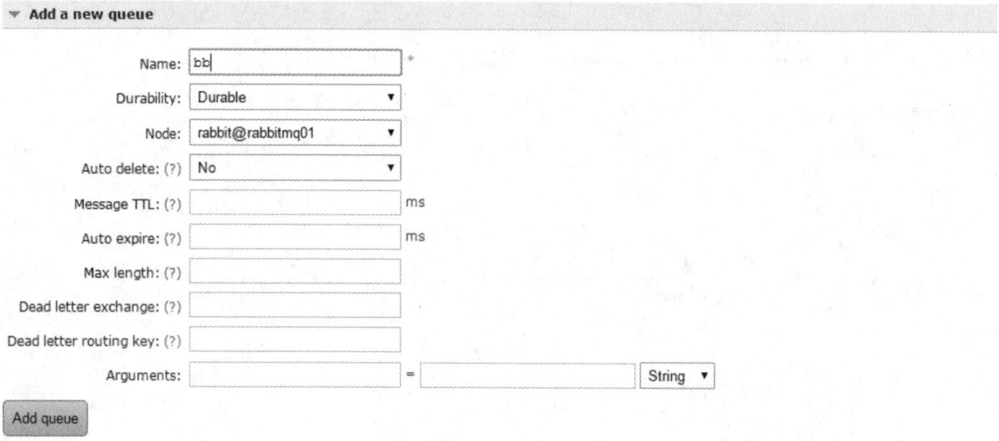

图 6.16　添加队列（3）

点击 Add queue 按钮，如图 6.17 所示。

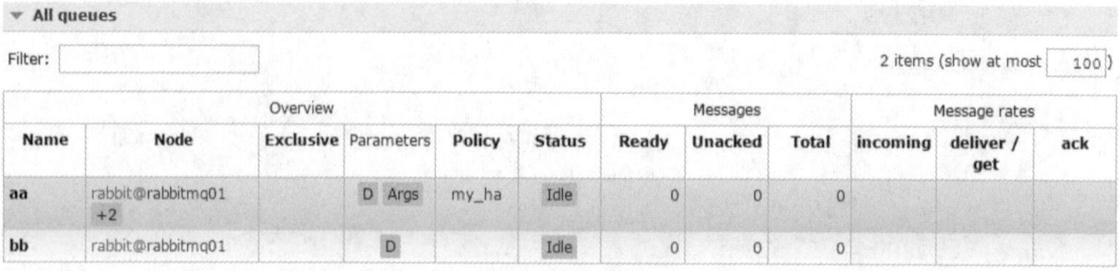

图 6.17　添加队列（4）

很明显我们没有看到 +2、Args、my_ha 这些参数。

最后分别在两个队列里创建一条消息，然后做一下破坏性测试看看集群效果是否成功？

点击 aa 队列按钮，拖动滚动条到最下面，然后创建消息，如图 6.18 所示。

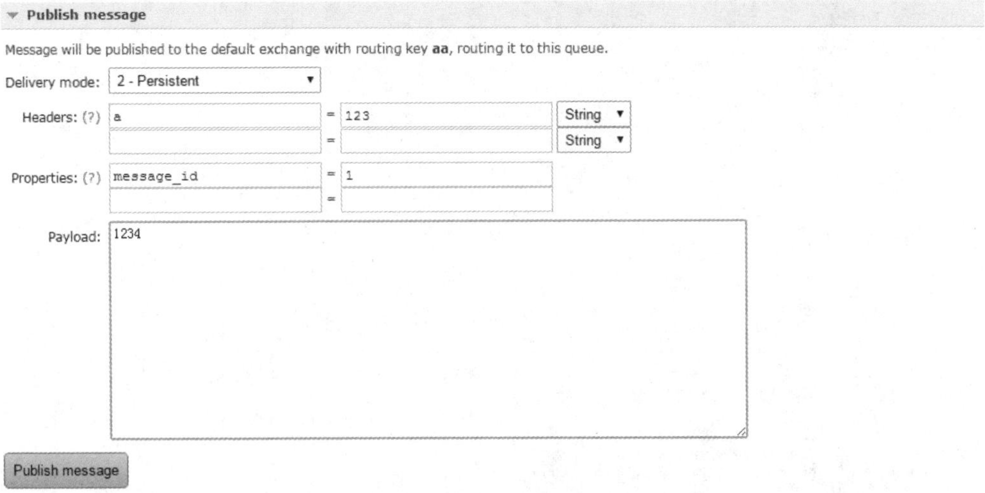

图 6.18 创建消息

2-Persistent 表示持久化，反之上面显示非持久化。

Headers：随便填写即可。

Properties：点击问号，选择一个消息 ID 号。

Payload：消息内容。

点击 Publish message 按钮，显示如图 6.19 所示。

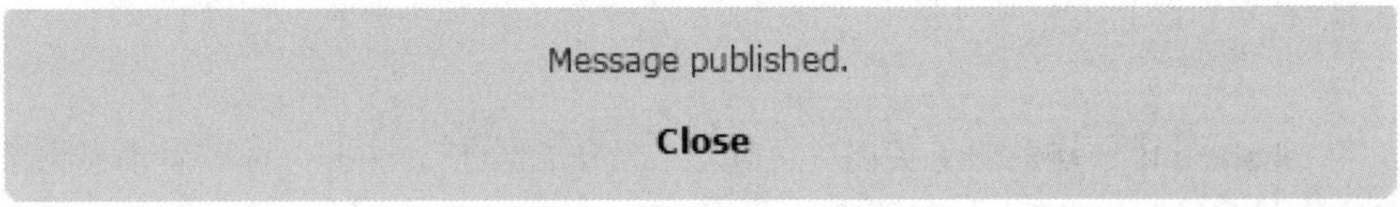

图 6.19 Publish message

点击 Close 按钮。然后再点击菜单栏的 Queues 按钮，可以看到 aa 队列的 Ready 和 Total 中多了一条消息记录，如图 6.20 所示。

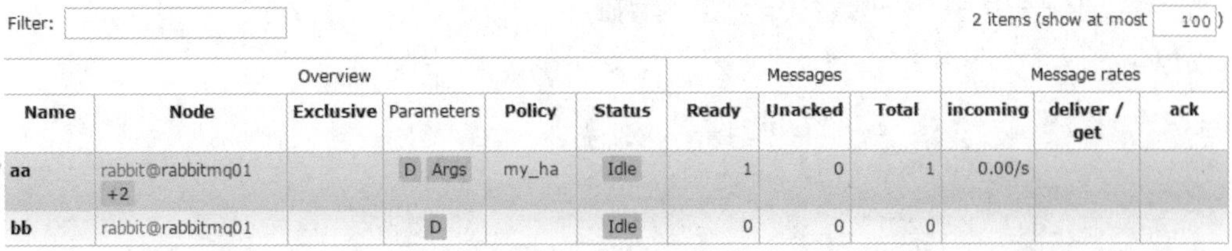

Filter:　　2 items (show at most 100)

Overview						Messages			Message rates		
Name	Node	Exclusive	Parameters	Policy	Status	Ready	Unacked	Total	incoming	deliver / get	ack
aa	rabbit@rabbitmq01 +2		D Args	my_ha	Idle	1	0	1	0.00/s		
bb	rabbit@rabbitmq01		D		Idle	0	0	0			

图 6.20 查看消息记录

然后再点击 bb 队列按钮，同样创建一条消息，设置同上，如图 6.21 所示。

点击 Publish message 按钮，同样还是点击 Queues 按钮，也看到 bb 队列里面也多了一条消息记录，如图 6.22 所示。

现在我们开始进行破坏性测试。

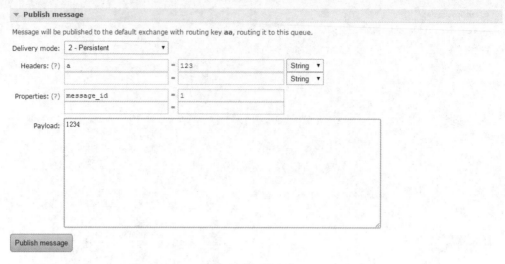

图 6.21　创建消息

All queues
Filter:
2 items (show at most 100)

Overview						Messages			Message rates		
Name	Node	Exclusive	Parameters	Policy	Status	Ready	Unacked	Total	incoming	deliver / get	ack
aa	rabbit@rabbitmq01 +2		D Args	my_ha	Idle	1	0	1	0.00/s		
bb	rabbit@rabbitmq01		D		Idle	1	0	1	0.00/s		

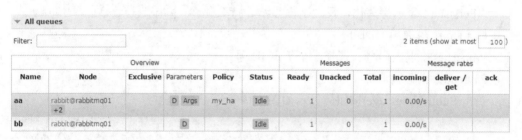

图 6.22　查看消息

将 rabbitmq01 节点的服务关闭，再通过 rabbitmq02 和 rabbitmq03 查看消息记录是否还存在，如图 6.23 和图 6.24 所示。

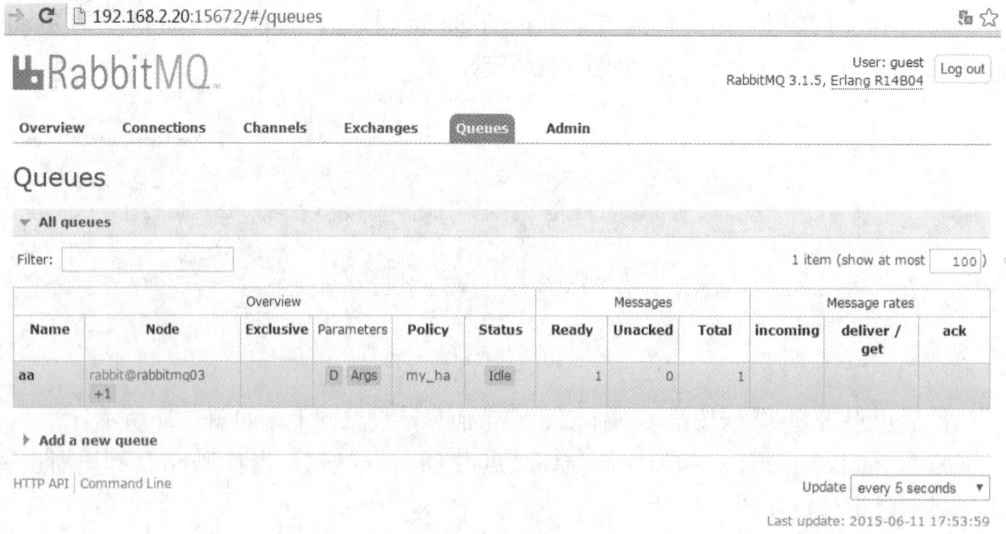

图 6.23　查看消息记录（1）

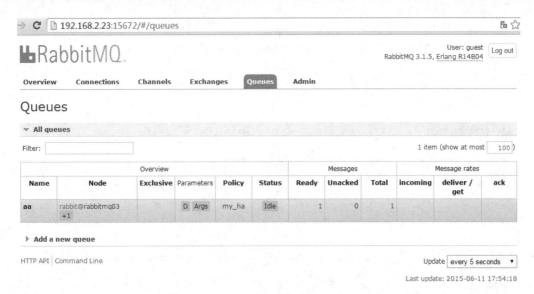

图 6.24　查看消息记录（2）

从上图可以看到 bb 队列已经不存在了，说明没有做镜像模式。而 aa 队列已经从之前的 +2 显示成 +1 了，而且消息记录还是存在的。

我们再将 rabbitmq02 节点的服务关闭，再通过 rabbitmq03 查看消息记录是否还存在，如图 6.25 所示。

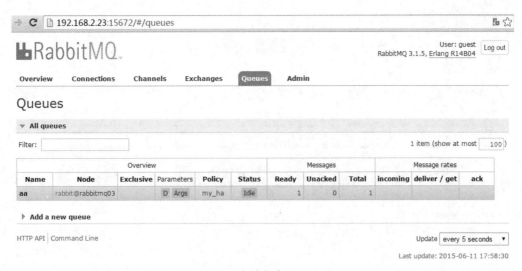

图 6.25　查看消息记录（3）

从上图可以看出 aa 消息队列和消息记录还是存在的，只是变成了一个节点了。我们再将 rabbitmq01 和 rabbitmq02 节点的服务启动起来，如图 6.26 所示。

bb 队列又成功回来了，但是这不是我们所关心的，因为它无法满足集群需要。我们看到了 aa 队列后面 +2 变成了粉色，鼠标指上去显示镜像无法同步。如果这个时候再将 rabbitmq03 节点的服务停掉，那么队列里面的消息将会丢失。手动在 rabbitmq01

或者是 rabbitmq02 节点上执行同步命令，aa 表示的是同步 aa 队列。下面我们选择 rabbitmq02 节点上执行 rabbitmqctl sync_queue aa，同步完成后 +2 又变成了蓝色，如图 6.27 所示。

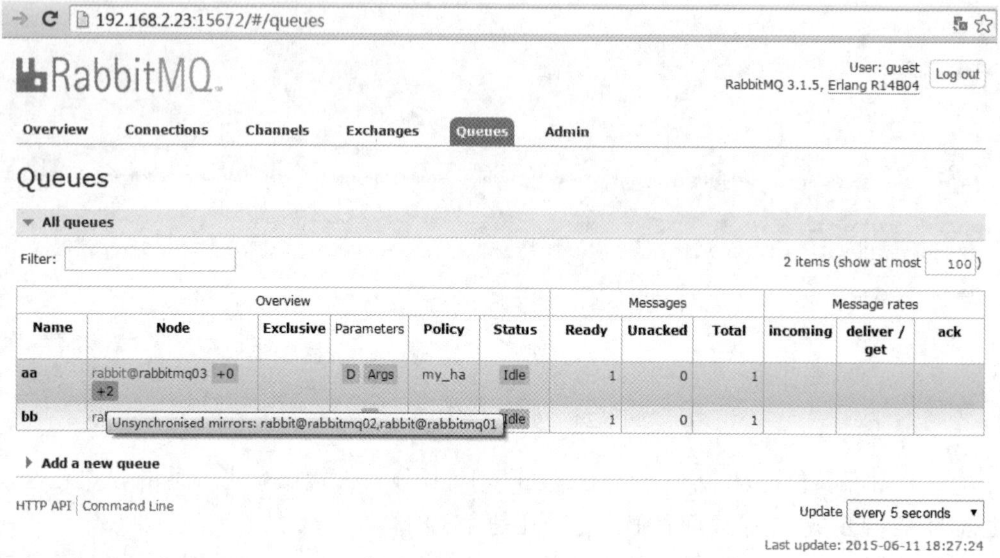

图 6.26 查看消息记录（4）

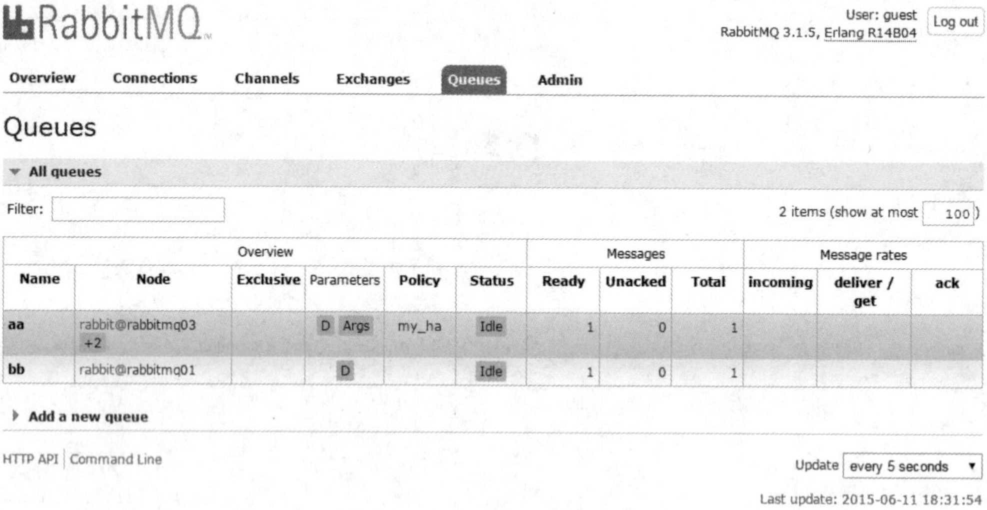

图 6.27 查看消息记录（5）

查看队列在服务器上的位置，如图 6.28 所示，一个目录对应的就是一个队列，里面就是该队列本身的消息。注意如果内存节点故障，队列就会丢失；如果是磁盘节点故障，队列还会存在。但如果是创建队列时指定了 ha 参数，待修复磁盘节点故障后，再启动磁盘节点的 rabbitmq 服务，未消费的队列及消息会不存在，需要手动执行同步队列命令。

```
[root@rabbitmq01 ~]# ll /var/lib/rabbitmq/mnesia/rabbit\@rabbitmq01/queues/
总用量 12
drwxr-xr-x 2 rabbitmq rabbitmq 4096 6月  12 08:30 1BL768WRH0RAX4M4MP8H16L5S
drwxr-xr-x 2 rabbitmq rabbitmq 4096 6月  12 08:30 81NI5XJ4KZDI2KGNS7MV3CP7T
```

图 6.28　查看队列在服务器上的位置

这样我们就测试了 rabbitmq 集群的破坏性测试，说明集群配置成功。但是消息节点一般都是程序直接调用，去创建队列、消息等等。程序里面如果想使用消息高可用集群配置，那么代码需要指定 ha 参数，如果不指定 x-ha-prolicy 的话将无法复制。

如图 6.29 所示是一段 C# 代码，仅供大家参考。

```csharp
using ( var bus = RabbitHutch.CreateBus(ConfigurationManager.ConnectionStrings["RabbitMQ"].ToString()))
    {
        bus.Subscribe< TestMessage>("word_subscriber" , message => RunTable(message),x=>x.WithArgument("x-ha-policy" , "all"));
        Console.WriteLine("Subscription Started. Hit any key quit" );
        Console.ReadKey();
    }
```

图 6.29　C# 代码

本章总结

- 大型网站技术架构的核心价值不是从无到有去搭建，而是能够伴随小型网站业务的逐步发展，慢慢地演化成一个大型网站。
- 从一开始一直需要做的是通过为用户提供好的服务来创造价值，得到用户的认可，提高用户忠诚度。
- 另外需要知道如何测试性能，当架构中有瓶颈时我们需要知道如何通过工具和软件找出瓶颈并去解决。

随手笔记

第 7 章

自动化运维之 Ansible

技能目标

- 理解 Ansible 核心概念
- 理解 Ansible 常见模块
- 会进行 Ansible 自动化操作
- 会进行 Ansible 自动化任务部署

本章导读

Ansible 基于 Python 开发,集合了众多优秀运维工具的优点,实现了批量运行命令、部署程序、配置系统等功能。默认通过 SSH 协议进行远程命令执行或下发配置,无需部署任何客户端代理软件,从而使得自动化环境部署变得更加简单。

知识服务

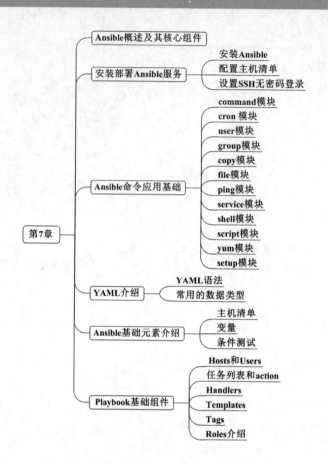

7.1　Ansible 概述

由于互联网的快速发展导致产品更新换代速度逐渐加快，运维人员每天都要进行大量的维护操作，仍旧按照传统方式进行维护会使得工作效率低下。这时，部署自动化运维就可以尽可能安全、高效地完成这些工作。

一般会把自动化运维工具划分为两类：一类是需要使用代理工具的，也就是基于专用的 Agent 程序来完成管理功能，如：Puppet、Func、Zabbix 等；另外一类是不需要配置代理工具的，可以直接基于 SSH 服务来完成管理功能，如：Ansible、Fabric 等。下面介绍几款功能类似的自动化运维工具。

1. Puppet

Puppet 基于 Ruby 开发，支持 Linux、UNIX、Windows 平台，可以针对用户、系统服务、配置文件、软件包等进行管理，有很强的扩展性，但远程执行命令相对较弱。

2. SaltStack

SaltStack 基于 Python 开发，允许管理员对多个操作系统创建统一的管理系统，比 Puppet 更轻量级。

3. Ansible

Ansible 基于 Python 开发，集合了众多优秀运维工具的优点，实现了批量运行命令、部署程序、配置系统等功能。默认通过 SSH 协议进行远程命令执行或下发配置，无需部署任何客户端代理软件，从而使得自动化环境部署变得更加简单。可同时支持多台主机并行管理，使得管理主机更加便捷。

针对这几款自动化运维工具的比较如表 7-1 所示。

表 7-1 自动化运维工具

工具	开发语言	结构	配置文件格式	运行任务
Ansible	Python	无	YAML	支持命令行
SaltStack	Python	C/S	YAML	支持命令行
Puppet	Ruby	C/S	Ruby 语法格式	通过模块实现

本章将主要介绍自动化运维工具 Ansible 是如何实现自动化运维部署的。

7.2 Ansible 核心组件

Ansible 可以看作是一种基于模块进行工作的框架结构，批量部署能力就是由 Ansible 所运行的模块实现的。简而言之 Ansible 是基于"模块"完成各种"任务"的。其基本框架结构如图 7.1 所示。

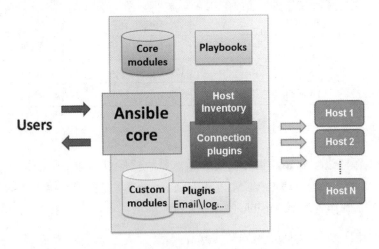

图 7.1 Ansible 基本框架结构

可以看出 Ansible 基本架构由六个部分构成：

Ansible core 核心引擎。

Host inventory 主机清单：用来定义 Ansible 所管理的主机，默认是在 Ansible 的

hosts 配置文件中定义被管理主机，同时也支持自定义动态主机清单和指定其他配置文件的位置。

Connection plugins 连接插件：负责和被管理主机实现通信。除支持使用 SSH 连接被管理主机外，Ansible 还支持其他的连接方式，所以需要有连接插件将各个主机用连接插件连接到 Ansible。

Playbooks（yaml，jinja2）剧本：用来集中定义 Ansible 任务的配置文件，即将多个任务定义在一个剧本中由 Ansible 自动执行，可以由控制主机针对多台被管理主机同时运行多个任务。

Core modules 核心模块：是 Ansible 自带的模块，使用这些模块将资源分发到被管理主机，使其执行特定任务或匹配特定的状态。

Custom modules 自定义模块：用于完成模块功能的补充，可借助相关插件完成记录日志、发送邮件等功能。

7.3 安装部署 Ansible 服务

Ansible 自动化运维环境由控制主机与被管理主机组成，由于 Ansible 是基于 SSH 协议进行通信的，所以控制主机安装 Ansible 软件后不需要重启或运行任何程序，被管理主机也不需要安装和运行任何代理程序。案例环境如表 7-2 所示。

表 7-2　Ansible 案例环境

角色	主机名	IP 地址	组名
控制主机	node1	192.168.46.167	
被管理主机	node2	192.168.46.168	websrvs
被管理主机	node3	192.168.46.169	websrvs
被管理主机	node4	192.168.46.170	dbsrvs

1. 安装 Ansible

可以使用源码进行安装，也可以使用操作系统软件包管理工具进行安装。这里使用 CentOS 7.2 操作系统，通过 YUM 方式安装 Ansible，需要依赖第三方的 EPEL 源，下面配置 EPEL 源作为部署 Ansible 的 YUM 源。

```
[root@node1 ~]# vi /etc/yum.repos.d/epel.repo
[epel]
name=epel
mirrorlist=http://mirrors.fedoraproject.org/mirrorlist?repo=epel-$releasever&arch=$basearch
enabled=1
gpgcheck=0
```

```
[root@node1 ~]# yum list ansible
Loaded plugins: fastestmirror, langpacks
Loading mirror speeds from cached hostfile
 * base: mirrors.aliyun.com
 * epel: mirrors.tuna.tsinghua.edu.cn
 * extras: mirrors.aliyun.com
Available Packages
ansible.noarch              2.2.0.0-4.el7              epel
```

准备好 YUM 源之后，就可以直接使用 yum 命令安装 Ansible。

```
[root@node1 ~]# yum install ansible
```

安装好后可以查看 Ansible 软件的版本信息。

```
[root@node1 ~]# ansible   --version
ansible 2.2.0.0
  config file = /etc/ansible/ansible.cfg
  configured module search path = Default w/o overrides
```

Ansible 主要相关配置文件在 /etc/ansible 目录下。

```
[root@node1 ansible]# pwd
/etc/ansible
[root@node1 ansible]# ll
total 24
-rw-r--r--. 1 root root 14396 Nov 11 22:37 ansible.cfg      // 配置文件
-rw-r--r--. 1 root root  1016 Nov 11 22:37 hosts            // 管控主机文件
drwxr-xr-x. 2 root root     6 Nov 11 22:37 roles
```

2. 配置主机清单

Ansible 通过读取默认主机清单 /etc/ansible/hosts 文件，修改主机与组配置后，可同时连接到多个被管理主机上执行任务。比如定义一个 websrvs 组，包含两台主机的 IP 地址，再定义一个 dbsrvs 组，包含一台主机的 IP 地址，内容如下：

```
[root@node1 ansible]# vi hosts
[websrvs]                        // 被管理主机分类
192.168.46.168
192.168.46.169
[dbsrvs]                         // 被管理主机分类
192.168.46.170
```

3. 设置 SSH 无密码登录

为了避免 Ansible 下发指令时输入被管理主机的密码，可以通过证书签名达到 SSH 无密码登录的效果，使用 ssh-keygen 产生一对密钥，使用 ssh-copy-id 来下发生成的公钥。

```
[root@node1 ~]#ssh-keygen -t rsa              // 基于 SSH 密钥的连接
[root@node1 ~]#ssh-copy-id  -i /root/.ssh/id_rsa.pub root@192.168.24.168
[root@node1 ~]#ssh-copy-id  -i /root/.ssh/id_rsa.pub root@192.168.24.169
[root@node1 ~]#ssh-copy-id  -i /root/.ssh/id_rsa.pub root@192.168.24.170
```

到此 Ansible 的环境就部署完成了。

7.4　Ansible 命令应用基础

Ansible 可以使用命令行方式进行自动化管理，基本语法如下：

```
ansible <host-pattern> [-m module_name] [-a args]
<host-pattern> 对哪些主机生效
[-m module_name] 要使用的模块
[-a args] 模块特有参数
```

Ansible 的命令行管理工具都是由一系列模块、参数所支持的，可以在命令后面加上 -h 或 --help 获取帮助。如使用 ansible-doc 工具可以通过 ansible-doc -h 或者 ansible-doc --help 查看其帮助信息。

ansible-doc 是用来查看模块帮助信息的工具，最主要的选项 -l 用来列出可使用的模块，-s 用来列出某个模块的描述信息和使用示列。如列出 yum 模块的描述信息和操作动作：

```
[root@node1 ~]# ansible-doc -s yum
- name: Manages packages with the 'yum' package manager
  action: yum
    conf_file  # The remote yum configuration file to use for the transaction.
    disable_gpg_check  #Whether to disable the GPG checking of signatures of packages being
        installed. Has an effect only if state is 'present' or 'latest'.
    ...
```

Ansible 自带了很多模块，能够下发执行 Ansible 的各种管理任务。首先来了解下 Ansible 常用的这些核心模块。

1. command 模块

Ansible 管理工具使用 -m 选项来指定使用模块，默认使用 command 模块，即 -m 选项省略时会运行此模块，用于在被管理主机上运行命令。

```
[root@node1 ~]# ansible-doc -s command
- name: Executes a command on a remote node
  action: command
  chdir   # cd into this directory before running command
  creates  #a filename or (since 2.0)glob pattern already exist this step will *not* be run.
  executable  # change the shell used to execute the command.Should be an
    absolute path to the executable.
    ……
```

例如，在被管理主机上执行 date 命令，显示被管理主机的时间。有三种执行命令的方式去管理写入主机清单中的主机。

（1）使用 IP 地址指定运行主机

```
[root@node1 ~]# ansible 192.168.46.168 -m command -a 'date'
192.168.46.168 | SUCCESS | rc=0 >>
Mon Nov 11 16:18:52 CST 2016
```

（2）使用被管理主机中的分类运行

```
[root@node1 ~]# ansible websrvs -m command -a 'date'
192.168.46.168 | SUCCESS | rc=0 >>
Mon Nov 11 16:22:25 CST 2016

192.168.46.169 | SUCCESS | rc=0 >>
Mon Nov 11 16:22:27 CST 2016

[root@node1 ~]# ansible dbsrvs -m command -a 'date'
192.168.46.170 | SUCCESS | rc=0 >>
Mon Nov 11 16:23:01 CST 2016
```

（3）在所有主机清单中的主机上运行

```
[root@node1 ~]# ansible all -m command -a 'date'
192.168.46.168 | SUCCESS | rc=0 >>
Mon Nov 11 16:23:30 CST 2016

192.168.46.170 | SUCCESS | rc=0 >>
Mon Nov 11 16:23:30 CST 2016

192.168.46.169 | SUCCESS | rc=0 >>
Mon Nov 11 16:23:30 CST 2016
```

若省略 -m 选项，默认运行 command 模块。

```
[root@node1 ~]# ansible all -a 'tail -1 /etc/passwd'
 192.168.46.169 | SUCCESS | rc=0 >>
oprofile:x:16:16:Special user account to be used by OProfile:/home/oprofile:/sbin/nologin

192.168.46.170 | SUCCESS | rc=0 >>
oprofile:x:16:16:Special user account to be used by OProfile:/home/oprofile:/sbin/nologin

192.168.46.168 | SUCCESS | rc=0 >>
oprofile:x:16:16:Special user account to be used by OProfile:/home/oprofile:/sbin/nologin
```

2. cron 模块

Ansible 中的 cron 模块用于定义任务计划。其中有两种状态（state）：present 表示添加（省略状态时默认使用），absent 表示移除。

```
[root@node1 ~]# ansible-doc -s cron
- name: Manage cron.d and crontab entries.
  action: cron
      backup    # If set, create a backup of the crontab before it is modified. The location of the
                backup is returned in the 'backup_file' variable by this module.
      cron_file # If specified, uses this file instead of an individual user's crontab. Ifthis is a relative
                path, it is interpreted with respect to /etc/cron.d. (If it is absolute, it will typically be /etc/
                crontab). To use the 'cron_file' parameter you must specify the 'user' as well.
      day       # Day of the month the job should run ( 1-31, *, */2, etc )
   hour   # Hour when the job should run ( 0-23, *, */2, etc )
   job    #The command to execute or, if env is set, the value of environmentvariable. Required if
          state=present.
   minute # Minute when the job should run ( 0-59, *, */2, etc )
   month  # Month of the year the job should run ( 1-12, *, */2, etc )
   state  # Whether to ensure the job or environment variable is present or absent.
   ……
```

（1）添加任务计划

```
[root@node1 ~]# ansible dbsrvs -m cron -a 'minute="*/10" job="/bin/echo hello" name="test cron job"'
192.168.46.170 | SUCCESS => {
    "changed": true,
    "envs": [],
    "jobs": [
        "test cron job"
    ]
}
[root@node1 ~]# ansible dbsrvs -a 'crontab -l'
192.168.46.170 | SUCCESS | rc=0 >>
#Ansible: test cron job
*/10 * * * * /bin/echo hello
```

（2）移除任务计划

```
[root@node1 ~]# ansible dbsrvs -m cron -a 'minute="*/10" job="/bin/echo hello"
name="test cron job" state=absent'
192.168.46.170 | SUCCESS => {
    "changed": true,
    "envs": [],
    "jobs": []
}
[root@node1 ~]# ansible dbsrvs -a 'crontab -l'
192.168.46.170 | SUCCESS | rc=0 >>
```

3. user 模块

Ansible 中的 user 模块用于创建新用户和更改、删除已存在的用户。其中 name 选项用来指明创建的用户名称。

（1）创建用户

```
[root@node1 ~]# ansible dbsrvs -m user -a 'name="user1"'
192.168.46.170 | SUCCESS => {
  "changed": true,
  "comment": "",
  "createhome": true,
  "group": 500,
  "home": "/home/user1",
  "name": "user1",
  "shell": "/bin/bash",
  "state": "present",
  "system": false,
  "uid": 500
}
```

（2）删除用户

```
[root@node1 ~]# ansible dbsrvs -m user -a 'name="user1" state=absent'
192.168.46.170 | SUCCESS => {
  "changed": true,
  "force": false,
  "name": "user1",
  "remove": false,
  "state": "absent"
}
```

4. group 模块

Ansible 中的 group 模块用于对用户组进行管理。

```
[root@node1 ~]# ansible-doc -s group
- name: Add or remove groups
  action: group
      gid      # Optional 'GID' to set for the group.
      name=    # Name of the group to manage.
      state    # Whether the group should be present or not on the remote host.
      system   # If 'yes', indicates that the group created is a system group.
```

例如，创建 mysql 组，将 mysql 用户添加到 mysql 组中。

```
[root@node1 ~]# ansible dbsrvs -m group -a 'name=mysql gid=306 system=yes'
192.168.46.170 | SUCCESS => {
  "changed": true,
  "gid": 306,
  "name": "mysql",
  "state": "present",
  "system": true
}
```

```
[root@node1 ~]# ansible dbsrvs -m user -a 'name=mysql uid=306 system=yes group=mysql'
192.168.46.170 | SUCCESS => {
  "changed": true,
  "comment": "",
  "createhome": true,
  "group": 306,
  "home": "/home/mysql",
  "name": "mysql",
  "shell": "/bin/bash",
  "state": "present",
  "system": true,
  "uid": 306
}
```

5. copy 模块

Ansible 中的 copy 模块用于实现文件复制和批量下发文件。其中使用 src 来定义本地源文件路径，使用 dest 定义被管理主机文件路径，使用 content 则是通过指定信息内容来生成目标文件。

```
[root@node1 ~]# ansible-doc -s copy
- name: Copies files to remote locations.
  action: copy
      backup  # Create a backup file including the timestamp information so you can get
                the original file back if you somehowclobbered it incorrectly.
      content  # When used instead of 'src', sets the contents of a file directly to the
specified value. This is for simplevalues, for anything complex or with
                formatting please switch to the templatemodule.
      dest=  # Remote absolute path where the file should be copied to. If src is adirectory, this must
           be a directory too.
      ......
```

例如，将本地文件 /etc/fstab 复制到被管理主机上的 /tmp/fstab.ansible，将所有者设置为 root，权限设置为 640。

```
[root@node1 ~]# ansible dbsrvs -m copy -a 'src=/etc/fstab dest=/tmp/fstab.ansible owner=root mode=640'
192.168.46.170 | SUCCESS => {
  "changed": true,
  "checksum": "d1f6ba55898cf7d7cbb19a9e1ed13f8e685825f7",
  "dest": "/tmp/fstab.ansible",
  "gid": 0,
  "group": "root",
  "md5sum": "f62196c2430b68cec3e015395198b56f",
  "mode": "0640",
  "owner": "root",
  "secontext": "unconfined_u:object_r:admin_home_t:s0",
  "size": 779,
```

```
    "src": "/root/.ansible/tmp/ansible-tmp-1475459696.6-226867162123326/source",
    "state": "file",
    "uid": 0
}

[root@node4 ~]# ll /tmp/fstab.ansible
-rw-r-----. 1 root root 779 Nov 11 22:04 /tmp/fstab.ansible
```

> **注意**
>
> 如果出现以下的报错信息，是因为被管理主机开启了 SELinux，需要在被管理机上安装 libselinux-python 软件包，才可以使用 Ansible 中与 copy、file 相关的函数。
>
> ```
> [root@node1 ~]# ansible dbsrvs -m copy -a 'src=/etc/fstab dest=/tmp/fstab.ansible owner=root mode=640'
> 192.168.46.170 | FAILED! => {
>     "changed": false,
>     "checksum": "d1f6ba55898cf7d7cbb19a9e1ed13f8e685825f7",
>     "failed": true,
>     "msg": "Aborting, target uses selinux but python bindings (libselinux-python)
>         aren't installed!"
> }
> ```

例如，将"Hello Ansible Hi Ansible"写入 /tmp/test.ansible 文件中。

```
[root@node1 ~]# ansible dbsrvs -m copy -a 'content="Hello Ansible Hi Ansible" dest=/tmp/test.ansible'
192.168.46.170 | SUCCESS => {
    "changed": true,
    "checksum": "9ab3d7aa8f7e110f060bdda8530775d75ac3a17e",
    "dest": "/tmp/test.ansible",
    "gid": 0,
    "group": "root",
    "md5sum": "4d14aa63dfe1b1a54a1390c6135b4ec2",
    "mode": "0644",
    "owner": "root",
    "secontext": "unconfined_u:object_r:admin_home_t:s0",
    "size": 25,
    "src": "/root/.ansible/tmp/ansible-tmp-1475460070.7-260004292343959/source",
    "state": "file",
    "uid": 0
}

[root@node4 ~]# ll /tmp/test.ansible
-rw-r--r--. 1 root root 24 Nov 11 22:09 /tmp/test.ansible
[root@node4 ~]# cat /tmp/test.ansible
Hello Ansible Hi Ansible
```

6. file 模块

在 Ansible 中使用 file 模块来设置文件属性。其中使用 path 指定文件路径，使用 src 定义源文件路径，使用 name 或 dest 来替换创建文件的符号链接。

```
[root@node1 ~]# ansible-doc -s file
- name: Sets attributes of files
  action: file
    follow  # This flag indicates that filesystem links, if they exist, should be followed.
    force   # force the creation of the symlinks in two cases: the source file doesnot exist (but
            will appear later); the destination exists and is a file (so, we need to unlink the "path" file
            and createsymlink to the "src" file in place of it).
    group   # name of the group that should own the file/directory, as would be fed to'chown'
    mode    # mode the file or directory should be. For those used to '/usr/bin/chmod'remember
            that modes are actually octal numbers (like 0644). Leaving off the leading zero will
            likely have unexpected results. As of version 1.8, the mode may be specified as a symbolic
            mode (for  example, 'u+rwx' or 'u=rw,g=r,o=r').
……
```

例如，设置文件 /tmp/fstab.ansible 的所属主为 mysql，所属组为 mysql，权限为 644。

```
[root@node1 ~]# ansible dbsrvs  -m file -a  'owner=mysql group=mysql mode=644 path=
 /tmp/fstab.ansible'
192.168.46.170 | SUCCESS => {
   "changed": true,
   "gid": 306,
   "group": "mysql",
   "mode": "0644",
   "owner": "mysql",
   "path": "/tmp/fstab.ansible",
   "secontext": "unconfined_u:object_r:admin_home_t:s0",
   "size": 779,
   "state": "file",
   "uid": 306
}

[root@node4 ~]# ll /tmp/fstab.ansible
-rw-r--r--. 1 mysql mysql 779 Nov 11 22:04 /tmp/fstab.ansible
```

例如，设置文件 /tmp/fstab.link 为文件 /tmp/fstab.ansible 的链接文件。

```
[root@node1 ~]# ansible dbsrvs  -m file -a  'path=/tmp/fstab.link src=/tmp/fstab.ansible state=link'
192.168.46.170 | SUCCESS => {
   "changed": true,
   "dest": "/tmp/fstab.link",
   "gid": 0,
   "group": "root",
   "mode": "0777",
   "owner": "root",
```

```
    "secontext": "unconfined_u:object_r:user_tmp_t:s0",
    "size": 18,
    "src": "/tmp/fstab.ansible",
    "state": "link",
    "uid": 0
}

[root@node4 ~]# ll /tmp/fstab.link
lrwxrwxrwx. 1 root root 18 Nov 11 22:25 /tmp/fstab.link -> /tmp/fstab.ansible
```

7. ping 模块

在 Ansible 中使用 ping 模块来检测指定主机的连通性。

```
[root@node1 ~]# ansible all -m ping
192.168.46.170 | SUCCESS => {
    "changed": false,
    "ping": "pong"
}
192.168.46.169 | SUCCESS => {
    "changed": false,
    "ping": "pong"
}
192.168.46.168 | SUCCESS => {
    "changed": false,
    "ping": "pong"
}
```

8. service 模块

在 Ansible 中使用 service 模块来控制管理服务的运行状态。其中，使用 enabled 表示是否开机自动启动，取值为 true 或者 false；使用 name 定义服务名称；使用 state 指定服务状态，取值分别为 started、stoped、restarted。

```
[root@node1 ~]# ansible-doc -s service
- name: Manage services.
  action: service
      arguments   # Additional arguments provided on the command line
      enabled     # Whether the service should start on boot. *At least one of state andenabled are
                  required.*
      name=       # Name of the service.
      pattern     # If the service does not respond to the status command, name a substringto look
                  for as would be found in theoutput of the 'ps' command as a stand-in for a status result.  If
                  the string is found, the service will be assumed to be running.
      runlevel    # For OpenRC init scripts (ex: Gentoo) only.  The runlevel that this service belongs to.
      sleep       # If the service is being 'restarted' then sleep this many seconds betweenthe stop
                  and start command. This helps to workaround badly behaving init scripts that exit
                  immediately after signaling a  process to stop.
```

state   # 'started'/'stopped' are idempotent actions that will not run commandsunless necessary. 'restarted' willalways bounce the service. 'reloaded' will always reload. *At least one of state and enabled are required.*

例如，查看 httpd 服务的状态。

[root@node1 ~]# ansible websrvs -a 'service httpd status'
192.168.46.168 | FAILED | rc=3 >>
httpd is stopped

192.168.46.169 | FAILED | rc=3 >>
httpd is stopped

[root@node1 ~]# ansible websrvs -a 'chkconfig --list httpd'
192.168.46.168 | SUCCESS | rc=0 >>
httpd   0:off   1:off   2:off   3:off   4:off   5:off   6:off

192.168.46.169 | SUCCESS | rc=0 >>
httpd   0:off   1:off   2:off   3:off   4:off   5:off   6:off

例如，启动 httpd 服务并设置为开机自动启动。

[root@node1 ~]# ansible websrvs -m service -a 'enabled=true name=httpd **state=started**'
192.168.46.169 | SUCCESS => {
   "changed": true,
   "enabled": true,
   "name": "httpd",
   "state": "started"
}
192.168.46.168 | SUCCESS => {
   "changed": true,
   "enabled": true,
   "name": "httpd",
   "state": "started"
}
[root@node1 ~]# ansible websrvs -a 'service httpd status'
192.168.46.169 | SUCCESS | rc=0 >>
httpd (pid 32156) is running...

192.168.46.168 | SUCCESS | rc=0 >>
httpd (pid 32202) is running...

[root@node1 ~]# ansible websrvs -a 'chkconfig --list httpd'
192.168.46.169 | SUCCESS | rc=0 >>
httpd   0:off   1:off   2:on   3:on   4:on   5:on   6:off

192.168.46.168 | SUCCESS | rc=0 >>
httpd   0:off   1:off   2:on   3:on   4:on   5:on   6:off

9. shell 模块

Ansible 中的 shell 模块可以在被管理主机上运行命令，并支持像管道符等功能的复杂命令。

```
[root@node1 ~]# ansible-doc -s shell
- name: Execute commands in nodes.
  action: shell
      chdir    # cd into this directory before running the command
      creates  # a filename, when it already exists, this step will *not* be run.
      executable  # change the shell used to execute the command. Should be an absolute path to
          the executable.
      free_form=  # The shell module takes a free form command to run, as a string. There's not an
          actual option named "free form". See the examples!
      removes  # a filename, when it does not exist, this step will *not* be run.
      warn # if command warnings are on in ansible.cfg, do not warn about thisparticular line if set
          to no/false.
```

例如，创建用户后使用无交互模式给用户设置密码。

```
[root@node1 ~]# ansible dbsrvs -m user -a 'name=user1'
192.168.46.170 | SUCCESS => {
    "changed": true,
    "comment": "",
    "createhome": true,
    "group": 500,
    "home": "/home/user1",
    "name": "user1",
    "shell": "/bin/bash",
    "state": "present",
    "stderr": "useradd: warning: the home directory already exists.\nNot copying any file from
        skel directory into it.\nCreating mailbox file: File exists\n",
    "system": false,
    "uid": 500
}

[root@node1 ~]# ansible dbsrvs -m shell -a 'echo redhat|passwd --stdin user1'
192.168.46.170 | SUCCESS | rc=0 >>
Changing password for user user1.
passwd: all authentication tokens updated successfully.
```

10. script 模块

Ansible 中的 script 模块可以将本地脚本复制到被管理主机上进行运行。需要注意的是，使用相对路径来指定脚本。

```
[root@node1 ~]# ansible-doc -s script
- name: Runs a local script on a remote node after transferring it
  action: script
```

```
creates  # a filename, when it already exists, this step will *not* be run.
free_form=  # path to the local script file followed by optional arguments.
removes  # a filename, when it does not exist, this step will *not* be run.
```

例如，编辑一个本地脚本 test.sh，复制到被管理主机上进行运行。

```
[root@node1 ~]# vi test.sh
#!/bin/bash
echo "hello ansible from script" > /tmp/script.ansible
[root@node1 ~]# chmod +x test.sh
[root@node1 ~]# ansible dbsrvs -m script -a 'test.sh'
192.168.46.170 | SUCCESS => {
   "changed": true,
   "rc": 0,
   "stderr": "",
   "stdout": "",
   "stdout_lines": []
}
[root@node4 ~]# cat /tmp/script.ansible
hello ansible from script
```

11．yum 模块

Ansible 中的 yum 模块负责在被管理主机上安装与卸载软件包，但是需要提前在每个节点配置自己的 YUM 仓库。其中使用 name 指定要安装的软件包，还需要带上软件包的版本号，否则安装最新的软件包；使用 state 指定安装软件包的状态，present、latest 用来表示安装，absent 表示卸载。

```
[root@node1 ~]# ansible-doc -s yum
- name: Manages packages with the 'yum' package manager
  action: yum
     conf_file  # The remote yum configuration file to use for the transaction.
     disable_gpg_check  #Whether to disable the GPG checking of signatures of packages being
            installed. Has an effect only if state is 'present' or 'latest'.
     disablerepo  # 'Repoid' of repositories to disable for the install/update operation. These repos
            will not persist beyond thetransaction. When specifying multiple repos, separate them
            with a ",".
     enablerepo  # 'Repoid' of repositories to enable for the install/update operation. These repos
            will not persist beyond the transaction. When specifying multiple repos, separate them
            with a ",".
     exclude  # Package name(s) to exclude when state=present, or latest
     list     # Various (non-idempotent) commands for usage with '/usr/bin/ansible' and'not'
            playbooks. See examples.
……
```

（1）安装 zsh 软件包

```
[root@node1 ~]# ansible dbsrvs -m yum -a 'name=zsh'
192.168.46.170 | SUCCESS => {
   "changed": true,
```

```
    "msg": "",
    "rc": 0,
    "results": [
      "Loaded plugins: fastestmirror, security\nLoading mirror speeds from cached hostfile\
       nSetting up Install Process\nResolving Dependencies\n--> Running transaction check\
       n---> Package zsh.x86_64 0:4.3.10-7.el6 will be installed\n--> Finished Dependency
       Resolution\n\nDependencies Resolved\n\n================================
       ==========================================\n Package         Arch
       Version        Repository      Size\n===============================
       ===========================================\nInstalling:\n zsh         x86_64
       4.3.10-7.el6       base       2.1 M\n\nTransaction Summary\n================
       =============================================================
       \nInstall    1 Package(s)\n\nTotal download size: 2.1 M\nInstalled size: 4.8 M\
       nDownloading Packages:\nRunning rpm_check_debug\nRunning Transaction Test\
       nTransaction Test Succeeded\nRunning Transaction\n\r  Installing : zsh-4.3.10-7.el6.x8
       6_64                    1/1 \n\r  Verifying  : zsh-4.3.10-7.el6.x86_64
       1/1 \n\nInstalled:\n  zsh.x86_64 0:4.3.10-7.el6                            \n\nComplete!\n"
    ]
}

[root@node4 ~]# rpm -q zsh
zsh-4.3.10-7.el6.x86_64
```

（2）卸载 zsh 软件包

```
[root@node1 ~]# ansible dbsrvs -m yum -a 'name=zsh state=absent'
192.168.46.170 | SUCCESS => {
    "changed": true,
    "msg": "",
    "rc": 0,
    "results": [
      "Loaded plugins: fastestmirror, security\nSetting up Remove Process\nResolving
       Dependencies\n--> Running transaction check\n---> Package zsh.x86_64 0:4.3.10-7.el6 will
       be erased\n--> Finished Dependency Resolution\n\nDependencies Resolved\n\n=======
       =================================================================
       =======\n Package      Arch        Version         Repository      Size\n=======
       =================================================================
       =======\nRemoving:\n zsh        x86_64       4.3.10-7.el6       @base      4.8
       M\n\nTransaction Summary\n=====================================================
       =====================\nRemove    1 Package(s)\n\
       nInstalled size: 4.8 M\nDownloading Packages:\nRunning rpm_check_debug\nRunning
       Transaction Test\nTransaction Test Succeeded\nRunning Transaction\n\r  Erasing    :
       zsh-4.3.10-7.el6.x86_64                        1/1 \n\r  Verifying  : zsh-4.3.10-7.el6.x86_64
       1/1 \n\nRemoved:\n  zsh.x86_64 0:4.3.10-7.el6                              \n\nComplete!\n"
    ]
}

[root@node4 ~]# rpm -q zsh
package zsh is not installed
```

12. setup 模块

在 Ansible 中使用 setup 模块收集、查看被管理主机的 facts（facts 是 Ansible 采集被管理主机设备信息的一个功能）。每个被管理主机在接收并运行管理命令之前，都会将自己的相关信息（操作系统版本、IP 地址等）发送给控制主机。

```
[root@node1 ~]# ansible-doc -s setup
- name: Gathers facts about remote hosts
  action: setup
     fact_path     # path used for local ansible facts (*.fact) - files in this dir will be run (if executable)
                    and their results be added to ansible_local facts if a fileis not executable it is read. Check
                    notes for Windows options. (from 2.1 on) File/results format can be json or ini-format
     filter        # if supplied, only return facts that match this shell-style (fnmatch) wildcard.
     gather_subset # if supplied, restrict the additional facts collected to the given subset.
……
```

例如，查看 dbsrvs 组的 facts 信息。

```
[root@node1 ~]# ansible dbsrvs -m setup        // 输出信息比较多，这里省略
```

7.5 YAML 介绍

YAML 是一种用来表达资料序列的格式，由于参考了其他多种语言，所以具有很高的可读性。YAML 是 YAML Ain't Markup Language 的缩写，即 YAML 不是 XML。不过在研发这种语言时，YAML 的意思其实是 Yet Another Markup Language（仍是一种标记语言）。其特性如下：

（1）具有很好的可读性，易于实现；
（2）表达能力强，扩展性好；
（3）和脚本语言的交互性好；
（4）有一个一致的信息模型；
（5）可以基于流来处理。

更多关于 YAML 的内容可以参见 http://www.yaml.org。

1. YAML 语法

YAML 的语法和其他语言类似，也可以表达散列表、标量等数据结构。其中结构（structure）通过空格来展示；序列（sequence）里的项用 "-" 来代表；Map 里的键值对用 ":" 来分隔。YAML 文件扩展名通常为：yaml，如：example.yaml。下面是 YAML 的一个示例：

```
name:John Smith
age: 41
gender: Male
```

```
        spouse:
        name: Jane Smith
        age: 37
        gender: Female
        children:
        - name: Jimmy Smith
              age: 17
              gender: Male
               - name: Jenny Smith
              age: 13
              gender: Female
```

2. 常用的数据类型

YAML 中有两种常用的数据类型，分别是 list 和 dictionary。

（1）list

列表（list）的所有元素均使用 "-" 开头，如：

```
-Apple
-Orange
-Strawberry
-Mango
```

（2）dictionary

字典（dictionary）通过 key 与 value 进行标识，如：

```
name: Example Developer
Job: Developer
Skill: Elite
```

也可以使用 key:value 的形式放置于 { } 中进行表示，如：

```
{ name: Example Developer, Job: Developer, Skill: Elite}
```

7.6　Ansible 基础元素介绍

1. Inventory（主机清单）

Ansible 为了更加便捷地管理主机，在主机清单中将被管理主机进行分组命名，默认的主机清单为 /etc/ansible/hosts 文件。主机清单可以设置为多个，也可以通过 Dynamic Inventory 动态生成。

Inventory 文件以中括号中的字符标识为组名，将主机分组管理，也可以将同一主机同时划分到多个不同的组中。如果被管理主机使用非默认的 SSH 端口，还可以在主机名之后用冒号加端口号的方式来进行标明，如：

```
[webservers]
www1.example.org
www2.example.org

[dbservers]
db1.example.org
db2.example.org:2222
```

如果被管理主机的主机名遵循类似的命名规则，就可以使用列表的方式标识各个主机，如：

```
[webservers]
www[01:05].example.org

[dbbservers]
db-[a:f].example.org
```

在 Inventory 中有几个重要的概念。

（1）主机变量

可以在定义主机时添加主机变量，以便在后续的 Playbook 中使用，如：

```
[webservers]
www1.magedu.com http_port=80 maxRequestsChild=808
www2.magedu.com http_port=8080 maxRequestsChild=909
```

（2）组变量

组变量是指给指定主机设置可以在 Playbook 中直接使用的变量，如：

```
[servers-vars]
ntp_server=ntp.example.org
nfs_server=nfs.example.org
```

（3）组嵌套

在 Inventory 中的组还可以嵌套其他的组，也可以向组中的主机指定变量。不过这些变量只能在 ansible-playbook 工具中使用，直接使用 Ansible 工具并不支持，如：

```
[apache]
httpd1.example.org
httpd2.example.org

[nginx]
ngx1.example.org
ngx2.example.org

[webservers:children]
apache
nginx
```

（4）Inventory 参数

Ansible 基于 SSH 连接 Inventory 中指定的被管理主机时，还可以通过参数指定交互方式，这些参数如表 7-3 所示。

表 7-3 Inventory 参数

参数	含义
ansible_ssh_port	指定 ssh 端口
ansible_ssh_user	指定 ssh 用户
ansible_ssh_pass	指定 ssh 用户登录是用认证密码，明文密码不安全
ansible_sudo_pass	指明 sudo 时候的密码
ansible_connection	SSH 连接的类型：local、ssh、paramiko
ansible_ssh_private_key_file	SSH 连接的公钥文件
ansible_shell_type	指定主机所使用的 shell 解释器，默认是 sh
ansible_python_interpreter	用来指定 python 解释器的路径
ansible_*_interpreter	用来指定主机上其他语法解释器的路径

如果不配置 SSH 密钥认证，就可以这样对被管理主机进行认证：

```
[root@node1 ~]# vim /etc/ansible/hosts
[websrvs]
192.168.46.168 ansible_ssh_user=root ansible_ssh_pass=redhat
```

2. 变量

在 Ansible 中变量名仅能由字母、数字和下划线组成，并且只能以字母开头。可以使用两种方式来传递 Ansible 变量。

（1）通过命令行传递变量

在运行 playbook 的时候，可以通过命令行的方式来传递一些变量提供给 playbook 使用，示例如下：

```
ansible-playbook test.yml –extra-vars "hosts-www user-mageedu"
```

（2）通过 roles 传递变量

当给一个主机应用角色（roles）的时候可以传递变量，然后在角色内使用这些变量，示例如下：

```
-hosts: webservers
 roles:
   -common
   -{role: foo_app_instance,dir:'/web/htdocs/a.com',port:8080}
```

3. 条件测试

如果需要根据变量、facts 或之前任务的执行结果来作为某 task 执行与否的前提时，就需要用到条件测试语句。

（1）when 语句

使用条件测试只需要在 task 之后添加 when 语句就可以，when 语句支持 jinja2 表达式语法，如：

```
tasks:
-name: "shutdown Debin flavored systems"
 command: /sbin/shutdown –h now
 when: ansible_os_family – "Debian"
```

when 语句中还可以使用 jinja2 的大多"filter"，例如要忽略此前某语句的错误并基于其结果（failed 或者 success）去运行后面指定的语句，可以使用类似示例如下：

```
tasks:
 - command: /bin/false
  register: result
  ignore_errors: True
 -command: /bin/something
  when: result|success
 -command: /bin/still/something_else
  when: result|skipped
```

此外，when 语句中还可以使用 facts 或 Playbook 中定义的变量。

条件测试的简单示例如下：

```
[root@node1 ~]# vim cond.yml
- hosts: all
 remote_user: root
 vars:
 - username: user10
 tasks:
 - name: create {{ username }}user
  user: name={{ username }}
  when: ansible_fqdn == "node4"
[root@node1 ~]# ansible-playbook cond.yml
PLAY [all] ****************************************************************

TASK [setup] **************************************************************
ok: [192.168.46.169]
ok: [192.168.46.170]
ok: [192.168.46.168]

TASK [create user10user] **************************************************
skipping: [192.168.46.168]
skipping: [192.168.46.169]
changed: [192.168.46.170]
```

```
PLA RECAP *********************************************************
192.168.46.168             : ok=1   changed=0   unreachable=0   failed=0
192.168.46.169             : ok=1   changed=0   unreachable=0   failed=0
192.168.46.170             : ok=2   changed=1   unreachable=0   failed=0
```

（2）迭代

当需要去执行重复任务时，可以使用迭代机制，直接将需要迭代的内容定义为 item 变量并进行引用，然后通过 with_items 语句来指明迭代的元素，如：

```
-name: add several users
 user: name={{ item }} state=present groups=wheel
  with_items:
    -testuser1
    -testuser2
```

从功能上来说，上面的语句等同于下面的语句：

```
-name: add user testuser1
 user: name=testuser1 state=present groups=wheel
-name: add user testuser2
 user: name= testuser2 state=present groups=wheel
```

定义循环列表 with_items 如下：

```
-apache
-php
-mysql-server
```

注意 with_items 中的列表值也可以是字典，但引用时需要使用 item.KEY 格式。

```
-{ name: apache, conf: conffiles/httpd.conf }
-{ name: php, conf: conffiles/php.ini }
-{ name: mysql-server, conf: conffiles/my.cnf }
```

事实上，在 with_items 中可以使用的元素还可为 hashes，如：

```
-name: add several users
 user: name={{ item.name }} state=present groups={{ item.groups }}
 with_items:
   -{ name: 'testuser1', groups:'wheel' }
   -{ name: 'testuser2', groups:'root' }
```

Ansible 的循环机制还有很多的功能，更具体的内容可以参考官方文档 :http://docs.ansible.com/playbook_loops.xml。

7.7 Playbook 介绍

Playbook 是由一个或多个 play 组成的列表，主要功能是将 task 定义好的角色归并为一组进行统一管理，也就是通过 task 调用 Ansible 的模板将多个 play 组织在一个 Playbook 中运行。

Playbooks 本身由以下各部分组成：

（1）Tasks：任务，即调用模块完成的某操作；

（2）Variables：变量；

（3）Templates：模板；

（4）Handlers：处理器，当某条件满足时，触发执行的操作；

（5）Roles：角色。

下面是一个 Playbook 的简单示例：

```
-hosts: webnodes                          // 定义的主机组，即应用的主机
 vars:                                    // 定义变量
  http_port:80
  max_clients:256
  remote_user:root
 tasks:                                   // 执行的任务
-name: ensure apache is at the lastest version
 yum: name=httpd state=started
 handlers:                                // 处理器
 -name: restart apache
  service: name=httpd state=restarted
```

接下来介绍 Playbook 的基础组件。

1. Hosts 和 Users 介绍

Playbook 的设计目的是为了让某个或某些主机以某个用户的身份去执行相应的任务。其中用于指定要执行指定任务的主机用 hosts 定义，可以是一个主机也可以是由冒号分隔的多个主机组；用于指定被管理主机上执行任务的用户用 remote_user 来定义，如上个示例中的：

```
-hosts:webnodes
 remote_user:root
```

remote_user 也可定义指定用户通过 sudo 的方法在被管理主机上运行指令，甚至可以在使用 sudo 时用 sudo_user 指定 sudo 切换的用户。

```
-hosts: webnodes
 remote_user: redhat
 tasks:
 -name: test connection
  ping:
  remote_user: redhat
  sudo: yes
```

2. 任务列表和 action 介绍

Play 的主体是任务列表（Tasks list）。任务列表中的任务按照次序逐个在 hosts 中

指定的所有主机上执行，在顺序执行这些任务时，如果发生错误会将所有已执行任务回滚，因此，需要在更正 Playbook 中的错误后重新执行这些任务。

task 的任务是按照指定的参数去执行模块。每个 task 都使用 name 输出 Playbook 的运行结果，一般输出内容为描述该任务执行的步骤，如果没有提供将输出 action 的运行结果。

定义 task 的格式可以用 action:module options 或 module:options，其中后者可以实现向后兼容。如果 action 的内容过多，可在行首使用空白字符进行换行。

```
tasks:
 -name: make sure apache is running
  service: name-httpd state=running
```

在 Ansible 自带模块中，command 模块和 shell 模块只需要一个列表定义即可，无需使用 key=value 格式，如：

```
tasks:
 -name: disable selinux
  command: /sbin/setenforce 0
```

如果由于命令或脚本运行结束使执行结果为零，可以使用如下代码替代：

```
tasks:
 -name: run this command and ignore the result
  Shell: /usr/bin/somecommand||/bin/true
```

或者使用 ignore_errors 来忽略错误信息：

```
tasks:
 -name: run this command and ignore the result
  shell: /usr/bin/somecommand
  ignore_errors: True
```

简单示例如下：

```
[root@node1 ~]# vim nginx.yml
-hosts: websrvs
 remote_user: root
 tasks:
 -name: create nginx group
  group: name=nginx system=yes gid=208
 -name: create nginx user
  user: name=nginx uid=208 group=nginx system=yes
-hosts: dbsrvs
 remote_user: root
 tasks:
 -name: copy file to dbsrvs
  copy: src=/etc/inittab dest=/tmp/inittab.ans

[root@node1 ~]# ansible-playbook nginx.yml
```

```
PLAY [websrvs] ************************************************************

TASK [setup] **************************************************************
ok: [192.168.46.168]
ok: [192.168.46.169]

TASK [create nginx group] *************************************************
changed: [192.168.46.168]
changed: [192.168.46.169]

TASK [create nginx user] **************************************************
changed: [192.168.46.168]
changed: [192.168.46.169]

PLAY [dbsrvs] *************************************************************

TASK [setup] **************************************************************
ok: [192.168.46.170]

TASK [copy file to dbsrvs] ************************************************
changed: [192.168.46.170]

PLAY RECAP ****************************************************************
192.168.46.168             : ok=3    changed=2    unreachable=0    failed=0
192.168.46.169             : ok=3    changed=2    unreachable=0    failed=0
192.168.46.170             : ok=2    changed=1    unreachable=0    failed=0
```

3. Handlers 介绍

Handlers 用于当关注的资源发生变化时所采取的操作。在 notify 中列出的操作便称为 handler，也就是在 notify 中需要调用 handler 中定义的操作。而 notify 这个动作在每个 play 的最后被触发，仅在所有的变化发生完成后一次性地执行指定操作。

```
-name: template configuration file
 template: src-template.j2 dest=/etc/foo.conf
 notify:
-restart memcached
-restart apache
```

handler 也是 task 列表的格式，如：

```
handlers:
  -name: restart memcached
service: name=memcached state=restarted
  -name: restart apache
    service: name=apache state=restarted
```

简单示例如下：

```
[root@node1 ~]# mkdir conf
[root@node1 ~]# cp /etc/httpd/conf/httpd.conf conf/
[root@node1 ~]# vi conf/httpd.conf
Listen 8080  // 修改端口号
[root@node1 ~]# vim apache.yml
- hosts: websrvs
  remote_user: root
  tasks:
   - name: install httpd package
     yum: name=httpd state=lastest
   - name: install configuration file or httpd
     copy: src=/root/conf/httpd.conf dest=/etc/httpd/conf/httpd.conf
   - name: start httpd service
     service: enabled=true name=httpd state=started
[root@node1 ~]# ansible websrvs -a 'service httpd stop'
192.168.46.169 | SUCCESS | rc=0 >>
Stopping httpd: [  OK  ]

192.168.46.168 | SUCCESS | rc=0 >>
Stopping httpd: [  OK  ]
```

确保 httpd 软件包没有安装：

```
[root@node1 ~]# ansible websrvs -m yum -a 'name=httpd state=absent'

[root@node1 ~]# ansible-playbook apache.yml
PLAY [websrvs] ************************************************************

TASK [setup] **************************************************************
ok: [192.168.46.169]
ok: [192.168.46.168]

TASK [install httpd package] **********************************************
changed: [192.168.46.168]
changed: [192.168.46.169]

TASK [install configuration file for httpd] *******************************
changed: [192.168.46.169]
changed: [192.168.46.168]

TASK [start httpd service] ************************************************
changed: [192.168.46.169]
changed: [192.168.46.168]

PLAY RECAP ****************************************************************
192.168.46.168             : ok=4    changed=3    unreachable=0    failed=0
```

```
192.168.46.169              : ok=4    changed=3    unreachable=0    failed=0
```

```
[root@node3 ~]# rpm -q httpd
httpd-2.2.15-29.el6.centos.x86_64
[root@node3 ~]# grep "Listen" /etc/httpd/conf/httpd.conf |grep -v "#"
Listen 8080
[root@node3 ~]# service httpd status
httpd (pid  35707) is running...
[root@node3 ~]# ss -tnlp
State     Recv-Q Send-Q  Local Address:Port     Peer Address:Port
……
LISTEN     0      128           :::8080            :::*     users:(("httpd",35707,5),("httpd",35711,
    5),("httpd",35712,5),("httpd",35713,5),("httpd",35714,5),("httpd",35715,5),("httpd",35716,
    5),("httpd",35717,5),("httpd",35718,5))
……
```

如果配置文件有改动，比如apache端口号改变，则需要定义notify和handlers，触发更新相关执行操作。

```
[root@node1 ~]# vim conf/httpd.conf
Listen 808
[root@node1 ~]# vim apache.yml
- hosts: websrvs
  remote_user: root
  tasks:
  - name: install httpd package
    yum: name=httpd state=latest
  - name: install configuration file for httpd
    copy: src=/root/conf/httpd.conf dest=/etc/httpd/conf/httpd.conf
    notify:
    -restart httpd
  - name: start httpd service
    service: enabled=true name=httpd state=started
  handlers:
  - name: restart httpd
    service: name=httpd state=restarted
[root@node1 ~]# ansible-playbook apache.yml
PLAY [websrvs] ************************************************************

TASK [setup] **************************************************************
ok: [192.168.46.169]
ok: [192.168.46.168]

TASK [install httpd package] **********************************************
ok: [192.168.46.169]
ok: [192.168.46.168]
```

TASK [install configuration file for httpd] *******************************
changed: [192.168.46.169]
changed: [192.168.46.168]

TASK [start httpd service] **
ok: [192.168.46.169]
ok: [192.168.46.168]

RUNNING HANDLER [restart httpd] ***
changed: [192.168.46.169]
changed: [192.168.46.168]

PLAY RECAP **
192.168.46.168   : ok=5   changed=2   unreachable=0   failed=0
192.168.46.169   : ok=5   changed=2   unreachable=0   failed=0

[root@node3 ~]# ss -tnlp
State   Recv-Q Send-Q   Local Address:Port   Peer Address:Port
……
LISTEN   0   128   :::808   :::*   users:(("httpd",37036,6),("httpd",37040,6),("httpd",37041,6),("httpd",37042,6),("httpd",37043,6),("httpd",37044,6),("httpd",37045,6),("httpd",37046,6),("httpd",37047,6))

如果引入变量，则以上 apache.yml 可以改写为：

[root@node1 ~]# vim apache.yml
- hosts: websrvs
  remote_user: root
  vars:
  -package: httpd
  -service: httpd
  tasks:
  - name: install httpd package
    yum: name={ {package} } state=latest
  - name: install configuration file for httpd
    copy: src=/root/conf/httpd.conf dest=/etc/httpd/conf/httpd.conf
    notify:
    -restart httpd
  - name: start httpd service
    service: enabled=true name={ { service } } state=started
  handlers:
  - name: restart httpd
    service: name={ { service } } state=restarted

也可以直接引用 Ansible 的变量，如：

[root@node1 ~]# vim test.yml
- hosts: websrvs

```
  remote_user: root
  tasks:
  - name: copy file
    copy: content="{{ ansible_all_ipv4_addresses }}" dest=/tmp/vars.ans
```

[root@node1 ~]# ansible-playbook test.yml
PLAY [websrvs] **

TASK [setup] **
ok: [192.168.46.169]
ok: [192.168.46.168]

TASK [copy file] **
changed: [192.168.46.168]
changed: [192.168.46.169]

PLAY RECAP **
192.168.46.168   : ok=2   changed=1   unreachable=0   failed=0
192.168.46.169   : ok=2   changed=1   unreachable=0   failed=0

[root@node3 ~]# cat /tmp/vars.ans
["192.168.46.169"]

引用 Ansible 主机变量，如：

[root@node1 ~]# vim /etc/ansible/hosts
[websrvs]
192.168.46.168 testvar="46.168"
192.168.46.169 testvar="46.169"
[dbsrvs]
192.168.46.170

[root@node1 ~]# vim test.yml
- hosts: websrvs
  remote_user: root
  tasks:
  - name: copy file
    copy: content="{{ ansible_all_ipv4_addresses }},{{ testvar }}" dest=/tmp/vars.ans

[root@node1 ~]# ansible-playbook test.yml
PLAY [websrvs] **

TASK [setup] **
ok: [192.168.46.169]
ok: [192.168.46.168]

TASK [copy file] **

```
changed: [192.168.46.168]
changed: [192.168.46.169]

PLAY RECAP ****************************************************************
192.168.46.168            : ok=2    changed=1    unreachable=0    failed=0
192.168.46.169            : ok=2    changed=1    unreachable=0    failed=0

[root@node2 ~]# cat /tmp/vars.ans
([u'192.168.46.168'], 46.167)
[root@node3 ~]# cat /tmp/vars.ans
([u'192.168.46.169'], 46.168)
```

4. Templates 介绍

Jinja 是基于 Python 的模板引擎。Template 类是 Jinja 的另一个重要组件，可以看作是一个编译过的模板文件，用来产生目标文本，传递 Python 的变量给模板去替换模板中的标记。

```
[root@node1 ~]# mkdir templates
[root@node1 ~]# cp conf/httpd.conf templates/httpd.conf.j2
[root@node1 ~]# vim templates/httpd.conf.j2
```

修改如下：

```
Listen {{ http_port }}
ServerName {{ ansible_fqdn }}
MaxClients    {{ maxClients }}
```

使用主机变量定义一个变量名相同，而值不同的变量，如：

```
[root@node1 ~]# vim /etc/ansible/hosts
[websrvs]
192.168.46.168  http_port=80 maxClients=100
192.168.46.169  http_port=8080 maxClients=200

[root@node1 ~]# vim apache.yml
- hosts: websrvs
  remote_user: root
  vars:
  - package: httpd
  - service: httpd
  tasks:
  - name: install httpd package
    yum: name={{ package }} state=latest
  - name: install configuration file for httpd
    template: src=/root/templates/httpd.conf.j2 dest=/etc/httpd/conf/httpd.conf
    notify:
    - restart httpd
```

```
  - name: start httpd service
    service: enabled=true name={{ service }} state=started
  handlers:
  - name: restart httpd
service: name={{ service }} state=restarted
```

[root@node1 ~]# ansible-playbook apache.yml
PLAY [websrvs] **

TASK [setup] **
ok: [192.168.46.168]
ok: [192.168.46.169]

TASK [install httpd package] **
ok: [192.168.46.168]
ok: [192.168.46.169]

TASK [install configuration file for httpd] ***********************************
changed: [192.168.46.169]
changed: [192.168.46.168]

TASK [start httpd service] **
ok: [192.168.46.169]
ok: [192.168.46.168]

RUNNING HANDLER [restart httpd] ***
changed: [192.168.46.168]
changed: [192.168.46.169]

PLAY RECAP **
192.168.46.168   : ok=5   changed=2   unreachable=0   failed=0
192.168.46.169   : ok=5   changed=2   unreachable=0   failed=0

[root@node2 ~]# grep -i listen /etc/httpd/conf/httpd.conf |grep -v "#"
Listen 80
grep -i servername /etc/httpd/conf/httpd.conf |grep -v "#"
ServerName node2

[root@node3 ~]# grep -i listen /etc/httpd/conf/httpd.conf |grep -v "#"
Listen 8080
[root@node3 ~]# grep -i servername /etc/httpd/conf/httpd.conf |grep -v "#"
ServerName node3

5. Tags 介绍

如果多次执行修改 Playbook 会涉及到一些没有变化的代码，可以使用 tags 让用户选择跳过没有变化的代码，只运行 Playbook 中发生变化的部分代码。可以在 Playbook

中为某个或某些任务定义"标签",在执行此 Playbook 时通过 ansible-playbook 命令使用 --tags 选项能实现仅运行指定的 tasks。

简单示例如下:

```
[root@node1 ~]# vim apache.yml
- hosts: websrvs
  remote_user: root
  vars:
  - package: httpd
  - service: httpd
  tasks:
  - name: install httpd package
    yum: name={{ package }} state=latest
  - name: install configuration file for httpd
    template: src=/root/templates/httpd.conf.j2 dest=/etc/httpd/conf/httpd.conf
    tags:
    - conf
    notify:
     - restart httpd
  - name: start httpd service
    service: enabled=true name={{ service }} state=started
  handlers:
  - name: restart httpd
    service: name={{ service }} state=restarted

[root@node1 ~]# vi /etc/ansible/hosts
[websrvs]
192.168.46.168  http_port=80 maxClients=150
192.168.46.169  http_port=8080 maxClients=250
```

只运行 Playbook 中的部分代码。

```
[root@node1 ~]# ansible-playbook apache.yml --tags="conf"
PLAY [websrvs] ****************************************************************

TASK [setup] ******************************************************************
ok: [192.168.46.169]
ok: [192.168.46.168]

TASK [install configuration file for httpd] ***********************************
changed: [192.168.46.168]
changed: [192.168.46.169]

RUNNING HANDLER [restart httpd] ***********************************************
changed: [192.168.46.169]
changed: [192.168.46.168]
```

```
PLAY RECAP *********************************************************************
192.168.46.168             : ok=3    changed=2    unreachable=0    failed=0
192.168.46.169             : ok=3    changed=2    unreachable=0    failed=0
```

特殊地，如果要始终在 Playbook 中运行某些代码，可以使用 tags: always 来进行标注，简单示例如下：

```
[root@node1 ~]# vim apache.yml
- hosts: websrvs
  remote_user: root
  vars:
  - package: httpd
  - service: httpd
  tasks:
  - name: install httpd package
    yum: name={{ package }} state=latest
    tags:
    - always
  - name: install configuration file for httpd
    template: src=/root/templates/httpd.conf.j2 dest=/etc/httpd/conf/httpd.conf
    tags:
    - conf
    notify:
     - restart httpd
  - name: start httpd service
    service: enabled=true name={{ service }} state=started
    tags:
    - service
  handlers:
  - name: restart httpd
    service: name={{ service }} state=restarted
```

6. Roles 介绍

Ansible 为了层次化、结构化地组织 Playbook，使用了角色（roles），可以根据层次结构自动装载变量文件、tasks 以及 handlers 等。只需要在 Playbook 中使用 include 指令即可使用 roles。简单来讲，roles 就是通过分别将变量、文件、任务、模块及处理器设置于单独的目录中，便捷地使用他们。

一个 roles 的案例如下：

```
site.yml                          // 主接口
webservers.yml
fooservers.yml
roles/
 common/
   files/
   templates/
   tasks/
```

```
    handlers/
    vars/
    meta/
    webservers/
     files/
     templates/
     tasks/
     handlers/
     vars/
     meta/
```

而在 Playbook 中，可以这样使用 roles，如：

```
-hosts: webservers
 roles:
  -common
  -webservers
```

也可以向 roles 传递参数，如：

```
-hosts: webservers
 roles:
  -common
  -{ role: foo_app_instance, dir: '/opt/a', port: 5000 }
  -{ role: foo_app_instance, dir: '/opt/b', port: 5001 }
```

甚至也可以条件式地使用 roles，如：

```
-hosts: webservers
 roles:
  -{ role: some_role, when: "ansible_os_family == 'RedHat' " }
```

创建 roles 时一般需要以下步骤：首先需要创建以 roles 命名的目录，然后在 roles 目录中分别创建以各角色名称命名的目录，如 webservers 等，在每个角色命名的目录中分别创建 files、handlers、meta、tasks、templates 和 vars 目录，用不到的目录可以创建为空目录，也可以不创建，最后在 Playbook 文件中调用各角色进行使用。

Roles 中各个目录中涉及的文件归纳如下：

（1）tasks 目录：至少应该包含一个名为 main.yml 的文件，用来定义此角色的任务列表，此文件可以使用 include 包含其他的位于此目录中的 task 文件。

（2）files 目录：存放由 copy 或 script 等模块调用的文件。

（3）templates 目录：template 模块会自动在此目录中寻找 jinja2 模板文件。

（4）handlers 目录：此目录中应当包含一个 main。

（5）yml 文件：用于定义此角色用到的各 handlers，在 handlers 中使用 include 包含的其他 handlers 文件也应该位于此目录中。

（6）vars 目录：应当包含一个 main.yml 文件，用于定义此角色用到的变量。

（7）meta 目录：应当包含一个 main.yml 文件，用于定义此角色的特殊设定及其依赖关系，只有 Ansible1.3 及其后的版本才支持。

（8）default 目录：为当前角色定义默认变量时使用此目录；应当包含一个 main.yml 文件。

Roles 的简单示例如下：

```
[root@node1 ~]# mkdir -p ansible_playbooks/roles/{websrvs,dbsrvs}/{tasks,files,templates.meta,handlers,vars}

[root@node1 ~]# tree ansible_playbooks/
ansible_playbooks/
└── roles
    ├── dbsrvs
    │   ├── files
    │   ├── handlers
    │   ├── tasks
    │   ├── templates.meta
    │   └── vars
    └── websrvs
        ├── files
        ├── handlers
        ├── tasks
        ├── templates.meta
        └── vars

13 directories, 0 files

[root@node1 ~]# cd ansible_playbooks/roles/websrvs/
[root@node1 websrvs]# ls
files  handlers  tasks  templates.meta  vars
[root@node1 websrvs]# cp /etc/httpd/conf/httpd.conf    files/
[root@node1 websrvs]# vim tasks/main.yml        //定义所有的任务
- name: install httpd package
  yum: name=httpd
- name: install configuration file
  copy: src=httpd.conf dest=/etc/httpd/conf/httpd.conf
  tags:
  - conf
  notify:
  - restart httpd
- name: start httpd
  service: name=httpd state=started
[root@node1 websrvs]# vim handlers/main.yml
- name:   restart httpd
  service: name=httpd state=restarted
```

```
[root@node1 websrvs]# cd /root/ansible_playbooks/
[root@node1 ansible_playbooks]# ls
roles
[root@node1 ansible_playbooks]# vim site.yml
- hosts: websrvs
  remote_user: root
  roles:
  - websrvs

[root@node1 ansible_playbooks]# ansible-playbook site.yml
PLAY [websrvs] **********************************************************

TASK [setup] ************************************************************
ok: [192.168.46.169]
ok: [192.168.46.168]

TASK [websrvs : install httpd package] **********************************
ok: [192.168.46.168]
ok: [192.168.46.169]

TASK [websrvs : install configuration file] *****************************
changed: [192.168.46.168]
changed: [192.168.46.169]

TASK [websrvs : start httpd] ********************************************
ok: [192.168.46.169]
ok: [192.168.46.168]

RUNNING HANDLER [websrvs : restart httpd] *******************************
changed: [192.168.46.169]
changed: [192.168.46.168]

PLAY RECAP **************************************************************
192.168.46.168             : ok=5    changed=2    unreachable=0    failed=0
192.168.46.169             : ok=5    changed=2    unreachable=0    failed=0
```

为了更加直观地修改 site.yml，针对不同主机去调用不同的角色。

```
[root@node1 ansible_playbooks]# pwd
/root/ansible_playbooks
[root@node1 ansible_playbooks]# vim site.yml
- hosts: 192.168.46.168
  remote_user: root
  roles:
```

```
  - websrvs
- hosts: 192.168.46.169
  remote_user: root
  roles:
  - dbsrvs
- hosts: 192.168.46.170
  remote_user: root
  roles:
  - websrvs
  - dbsrvs
```

部署 dbsrvs 相关如下：

```
[root@node1 ansible_playbooks]# cd roles/dbsrvs/
[root@node1 dbsrvs]# cp /etc/my.cnf files/
[root@node1 dbsrvs]# vim tasks/main.yml
- name: install mysql-server package
  yum name=mysql-server state=latest
- name: install configuration file
  copy: src=my.cnf dest=/etc/my.cnf
  tags:
  - myconf
  notify:
  - restart mysqld
- name: start mysql-service
  service: name=mysqld enabled=true state=started
[root@node1 dbsrvs]# vim handlers/main.yml
- name: restart mysqld
  service: name=mysqld state=restarted
```

重新运行 ansible-playbook 如下：

```
[root@node1 ansible_playbooks]# pwd
/root/ansible_playbooks
[root@node1 ansible_playbooks]# vim roles/dbsrvs/tasks/main.yml
[root@node1 ansible_playbooks]# vim roles/dbsrvs/tasks/main.yml
[root@node1 ansible_playbooks]# ansible-playbook site.yml
PLAY [192.168.46.168] ****************************************************

TASK [setup] *************************************************************
ok: [192.168.46.168]

TASK [websrvs : install httpd package] ***********************************
ok: [192.168.46.168]
```

TASK [websrvs : install configuration file] ************************************
ok: [192.168.46.168]

TASK [websrvs : start httpd] ***
ok: [192.168.46.168]

PLAY [192.168.46.169] **

TASK [setup] ***
ok: [192.168.46.169]

TASK [dbsrvs : install mysql-server package] ***********************************
changed: [192.168.46.169]

TASK [dbsrvs : install configuration file] *************************************
ok: [192.168.46.169]

TASK [dbsrvs : start mysql-service] **
changed: [192.168.46.169]

PLAY [192.168.46.170] **

TASK [setup] ***
ok: [192.168.46.170]

TASK [websrvs : install httpd package] ***
ok: [192.168.46.170]

TASK [websrvs : install configuration file] ************************************
ok: [192.168.46.170]

TASK [websrvs : start httpd] ***
changed: [192.168.46.170]

TASK [dbsrvs : install mysql-server package] ***********************************
changed: [192.168.46.170]

TASK [dbsrvs : install configuration file] *************************************
ok: [192.168.46.170]

TASK [dbsrvs : start mysql-service] **
changed: [192.168.46.170]

PLAY RECAP ***

192.168.46.168	: ok=4	changed=0	unreachable=0	failed=0
192.168.46.169	: ok=4	changed=2	unreachable=0	failed=0
192.168.46.170	: ok=7	changed=3	unreachable=0	failed=0

下面是创建 roles 时的注意事项：

（1）目录名同角色名的定义。

（2）目录结构有固定格式：

1）files：静态文件；

2）templates：Jinja2 模板文件；

3）tasks：至少有 main.yml 文件，定义各 tasks；

4）handlers：至少有一个 main.yml 文件，定义各 handlers；

5）vars：至少有一个 main.yml，定义变量；

6）meta：定义依赖关系等信息。

（3）在 roles 之外，通过 site.yml 定义 Playbook，额外也可以有其他的 yml。

本章总结

- Ansible 是一款通过 SSH 协议就可以远程执行或下发配置的自动化环境部署软件。
- Ansible 架构包括 Ansible core 核心引擎、Host inventory 主机清单、Connect plugin 连接插件、Playbook 剧本、Core modules 核心模块，Custom modules 自定义模块这几部分组成。
- 在使用 Ansible 部署自动化运维环境时，需注意以下几点：使用 Inventory 集中管理应用部署在哪些主机上；使用 Modules 设置调用哪些模块，使用 Ad-Hoc Commands 来指定运行命令；使用 Playbooks 进行复杂应用的部署。

第 8 章

自动化运维之 SaltStack

技能目标

- 掌握 SaltStack 的原理和安装
- 掌握 SaltStack 的常用模块
- 掌握 SaltStack 的 grains、pillar、state

本章导读

在生产环境中,服务器往往不只一台,有可能是成千上万台。对于运维人员来说,如果单独对每台服务器进行管理,工作难度实在是太大了。SaltStack 是一个服务器基础设施管理工具,它具有配置管理、远程执行、监控等功能。SaltStack 由 Python 语言编写,是非常简单易用和轻量级的管理工具。

知识服务

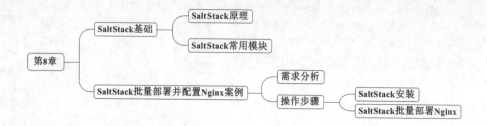

8.1　SaltStack 基础

通过部署 Saltstack 环境，可以在成千上万台服务器上批量执行命令。对于不同的业务进行集中管理、分发文件、采集数据、软件包管理等，有利于运维人员提高工作效率，规范业务配置和操作。

1. SaltStack 原理

SaltStack 由 Master 和 Minion 构成，Master 是服务端，表示一台服务器；Minion 是客户服务端，表示多台服务器。在 Master 上发送命令给符合条件的 Minion，Minion 就会执行相应的命令，Master 和 Minion 之间是通过 ZeroMQ（消息队列）进行通信的。

SaltStack 的 Master 端监听 4505 与 4506 端口，4505 为 Master 和 Minion 认证通信端口，4506 为 Master 用来发送命令或者接收 Minion 的命令执行返回信息。

当客户端启动后，会主动连接 Master 端注册，然后一直保持该 TCP 连接，而 Master 通过这条 TCP 连接对客户端进行控制。如果连接断开，Master 对客户端将不能进行控制。但是，当客户端检查到连接断开后，会定期向 Master 端请求注册连接。

2. SaltStack 常用模块

SaltStack 提供了非常多的功能模块，以便于对操作系统的基础功能和常用工具的操作。

（1）pkg 模块

pkg 模块的作用是包管理，包括增删更新。

（2）file 模块

file 模块的作用是管理文件操作，包括同步文件、设置文件权限和所属用户组、删除文件等操作。

（3）cmd 模块

cmd 模块的作用是在 Minion 上执行命令或者脚本。

（4）user 模块

user 模块的作用是管理系统账户操作。

（5）service 模块

service 模块的作用是管理系统服务操作。

（6）cron 模块

cron 模块的作用是管理 cron 服务操作。

关于 SaltStack 的详细介绍请访问课工场网站。

8.2 SaltStack 批量部署并配置 Nginx

8.2.1 需求分析

1. 案例目的

在生产环境中，经常需要根据不同的业务需求来分组部署和配置 Nginx 服务器。本案例使用了三台服务器，均采用 CentOS6.3 系统版本且最小化安装，要求能连接互联网，SELinux 和防火墙均已关闭。

需要注意的是，三台服务器一定要设置完整的 FQDN，和域名一样的形式，不然在主控端执行远程执行命令或者配置的时候，等待的时间会非常长，甚至还会出现其他不可控的问题。

2. 部署环境

三台服务器的部署参数如表 8-1 所示。

表 8-1 服务器参数

角色	主机名	IP 地址	组名	cpu 个数	Nginx 根目录
master	master.saltstack.com	192.168.85.135	—	—	
minion	web01.saltstack.com	192.168.85.141	web01group	1	/data
minion	web02.saltstack.com	192.168.85.142	web02group	2	/www

8.2.2 操作步骤

1. SaltStack 安装

下面演示部署服务器，步骤如下：

（1）首先三台服务器上都需要安装 epel 源。因为后面需要安装 SaltStack 服务端和客户端，也包括后面的 Nginx，如下所示：

```
[root@www ~]# rpm -ivh http://dl.fedoraproject.org/pub/epel/6/x86_64/epel-release-6-8.noarch.rpm
Retrieving http://dl.fedoraproject.org/pub/epel/6/x86_64/epel-release-6-8.noarch.rpm
warning: /var/tmp/rpm-tmp.HO5Bjk: Header V3 RSA/SHA256 Signature, key ID 0608b895: NOKEY
Preparing...                ########################################### [100%]
   1:epel-release           ########################################### [100%]
```

（2）对 /etc/yum.repos.d/epel.repo 做如下修改：

```
#baseurl
mirrorlist
改成
baseurl
#mirrorlist
```

在主控端（也就是 master）上安装 SaltStack 软件，命令如下：

```
yum install salt-master -y
```

在被控端（也就是两台 minion）上安装 SaltStack 软件，命令如下：

```
yum install salt-minion –y
```

（3）配置主控端配置文件 /etc/salt/master，注意默认的 master 文件全部是注释的。修改第 15 行的监听地址，注意为了安全，监听的地址一定要写内网地址。

```
interface:192.168.85.135
```

修改第 215 行的主控端会自动认证被控端的认证，只要被控端在设置完主控端的 IP 地址后启动服务，主控端就会允许被控端自动认证，以避免以后每次都要运行 salt-key 来确认证书信任。

```
auto_accept: True
```

修改第 416 行 saltstack 文件根目录位置，注意这个目录默认是没有的，需要创建。

```
file_roots:
  base:
    - /srv/salt
```

修改第 706 行的组分类。

```
nodegroups:
  web01group: 'web01.saltstack.com'
  web02group: 'web02.saltstack.com'
```

修改第 552 行的 pillar 开启功能。

```
pillar_opts: True
```

修改第 529 行的 pillar 的主目录，注意这个目录默认是没有的，需要创建。

```
pillar_roots:
  base:
    - /srv/pillar
```

主控端主要修改了以下内容：

```
[root@master ~]# cat /etc/salt/master | grep -v ^$ | grep -v ^#
interface:192.168.85.135
```

```
auto_accept:True
file_roots:
  base:
    - /srv/salt
nodegroups:
  web01group: 'web01.saltstack.com'
  web02group: 'web02.saltstack.com'
pillar_opts: True
pillar_roots:
  base:
    - /srv/pillar
```

主控端做完上述操作后启动 salt-master 服务。

```
[root@master ~]# /etc/init.d/salt-master start
Starting salt-master daemon:
```

启动后监听 TCP4505、TCP4506 端口。生产环境如果启用防火墙，建议主控端开放这两个端口。

创建 salt 文件根目录及 pillar 目录。

```
[root@master ~]# mkdir /srv/salt
[root@master ~]# mkdir /srv/pillar
```

（4）配置两台被控端配置文件 /etc/salt/minion（注意默认的 minion 文件全部也是注释的）

```
master: 192.168.85.135          //16 行，指定主控端 IP
id: web01.saltstack.com         //72 行，指定被控主机名，另一台是 web02.saltstack.com
```

分别启动两台被控端服务。

```
[root@web01 ~]# /etc/init.d/salt-minion start
Starting salt-minion daemon:                    [ 确定 ]
```

在主控端上简单测试一下主控端和被控端的通信状态，如果返回都是 True，则说明正常。注意这里的 ping 和我们平时用的 ping 命令不同，它只是 test 类下面的一个方法而已，用来验证主控端和被控端的通信状态。注意 * 表示所有通过认证的被控端，还可以支持其他很多正则表达式的匹配。

```
[root@master ~]# salt '*' test.ping
web01.saltstack.com:
    True
web02.saltstack.com:
    True
```

2．SaltStack 批量部署 Nginx

下面我们主要利用 SaltStack 的几个组件完成 Nginx 的安装和配置工作。主要由

grains、pillar、state 三个重要的组件完成。grains 和 pillar 都是采集被控端数据的，但是 grains 的特性在每次启动后汇报，没有 pillar 灵活，要知道 pillar 是随时可变的，只要在 master 端修改了，一般都会立刻生效。所以 grains 更适合做一些静态的属性值的采集，例如设备的角色、磁盘个数等诸如此类非常固定的属性。那么我们就可以得到一个大致的判断，如果你想定义的属性值是经常变化的，就采用 pillar；如果很固定、不易变，就采用 grains。

在配置之前，首先我们拆分一下 Nginx 的主配置文件，要根据当前 CPU 个数设置一些值，比如一些打开文件句柄数等。

上面说到 grains 可以采集被控端主机的一些值，一般定义 grains 数据的方法有两种，其中一种为在被控端主机上定制配置文件，另外一种是通过主控端扩展模块 API 实现，区别是模块更灵活。下面我们用自定义的方式进行配置。

通过下面的命令可以查看被控机 web01 主机上的 grains 所有值。

```
[root@master ~]# salt 'web01.saltstack.com' grains.items
we01.saltstack.com:
    ----------
    SSDs:
    biosreleasedate:
        07/02/2015
    biosversion:
        6.00
    cpu_flags:
        - fpu
        - vme
        - de
        - pse
        - tsc
        - msr
        - pae
        - mce
        - cx8
        - apic
        - mtrr
        - pge
        - mca
```
// 省略内容

grains 是 SaltStack 的重要组件之一，可以收集被控主机的基本信息，如 cpu、内核、操作系统、虚拟化等静态数据。在服务端可以利用这些信息对不同被控主机进行个性化配置。

但是默认打开的文件句柄数 1024 在被控端没有，需要我们通过 Python 脚本自定义获取。

state 组件是 SaltStack 的核心功能，通过预先定制好的文件对被控端主机进行状态管理，支持包含程序包（pkg）、文件（file）、网络配置（network）、系统服务（service）、系统用户（user）等服务，也就是定义好相关动作让被控端主机去执行。

（1）首先创建 grains 目录，将目录下的定制文件同步到被控端主机上运行，然后能正常获取被控端主机打开文件句柄数。

```
[root@master ~]# mkdir /srv/salt/_grains
[root@master ~]# vi /srv/salt/_grains/nginx_config.py
#!/usr/bin/python
import os,sys,commands
def NginxGrains():
    '''
        return Nginx config grains value
    '''
    grains = {}
    max_open_file=65535

    try:
        getulimit = commands.getstatusoutput('source /etc/profile;ulimit -n')
    except Exception,e:
        pass
    if getulimit[0]==0:
        max_open_file = int(getulimit[1])
    grains['max_open_file'] = max_open_file
    return grains
```

上述脚本的含义就是让被控端主机获取它当前打开文件句柄数。

在同步到被控端之前我们先执行一下如下命令，确认在主控端是否能获取被控端的 max_open_file 值。

```
[root@master ~]# salt '*' grains.item max_open_file
web01.saltstack.com:
    ----------
    max_open_file:
web02.saltstack.com:
    ----------
    max_open_file:
```

从命令执行结果中，我们看到不能获取到 max_open_file 的值，那么需要同步 grains 模块，在主控端运行命令如下：

```
[root@master ~]# salt '*' saltutil.sync_all
web01.saltstack.com:
    ----------
    beacons:
    grains:
        - grains.nginx_config
    modules:
```

```
            output:
            renderers:
            returners:
            sdb:
            states:
            utils:
        web02.saltstack.com:
            ----------
            beacons:
            grains:
                - grains.nginx_config
            modules:
            output:
            renderers:
            returners:
            sdb:
            states:
            utils:
```

被控端文件存放在 /var/cache/salt 目录下。

```
[root@web01 ~]# find /var/cache/salt/
/var/cache/salt/
/var/cache/salt/minion
/var/cache/salt/minion/module_refresh
/var/cache/salt/minion/extmods
/var/cache/salt/minion/extmods/grains
/var/cache/salt/minion/extmods/grains/nginx_config.py
/var/cache/salt/minion/proc
/var/cache/salt/minion/files
/var/cache/salt/minion/files/base
/var/cache/salt/minion/files/base/_grains
/var/cache/salt/minion/files/base/_grains/nginx_config.py
```

再次来获取一下 max_open_file 的值，一会儿我们可以用 grains 获取到这个值，即配置 Nginx 主配置文件的每个线程打开的连接数。

```
[root@master ~]# salt '*' grains.item max_open_file
web01.saltstack.com:
    ----------
    max_open_file:
        1024
web02.saltstack.com:
    ----------
    max_open_file:
        1024
```

（2）配置 pillar，在主控端上创建入口文件 top.sls，入口文件的作用是定义 pillar

的数据覆盖被控主机的有效域范围，内容如下：

```
[root@master ~]# vi /srv/pillar/top.sls
base:
  web01group:                         // 组名
    - match: nodegroup
    - web01server                     // 指定包括 seb01server.sls
  web02group:                         // 组名
    - match: nodegroup
    - web02server                     // 指定包括 seb02server.sls
```

web01group 和 web02group 是 /etc/salt/master 中定义的不同的组，对每一个组编写一个对应的文件指定配置，这里使用的是 web01server 和 web02server，再分别定义不同组主机的 nginx 的根目录，如下所示：

```
[root@master ~]# vi /srv/pillar/web01server.sls
nginx:
  root: /data
[root@master ~]# vi /srv/pillar/web02server.sls
nginx:
  root: /www
```

使用以下命令查看 pillar 配置的情况。

```
[root@master ~]# salt '*' pillar.data nginx
we01.saltstack.com:
    ----------
    nginx:
      ----------
      root:
        /data
we02saltstack.com:
    ----------
    nginx:
      ----------
      root:
        /www
```

从执行结果中可以很明显地看出，被控主机的 nginx 的根目录被配置成功。

（3）配置 state，state 是 SaltStack 最核心的功能，通过预先定制好的 sls 文件（salt state file），对被控主机进行管理，如程序包、文件、网络配置、系统服务、系统用户等。state 的定义也是通过编写 sls 文件进行操作，首先定义 state 的入口 top.sls 文件，注意和 pillar 的入口文件名字一样，内容如下：

```
[root@master ~]# vi /srv/salt/top.sls
base:
  '*':
    - nginx
```

其次定义被控机执行的状态，安装 nginx 软件，配置文件并启动。

```
[root@master ~]# vi /srv/salt/nginx.sls
nginx:
  pkg:                                      //1. 包管理
    - installed                             // 安装包

  file.managed:                             //2. 文件管理
    - source: salt://nginx/nginx.conf       // 配置文件在服务器路径
    - name: /etc/nginx/nginx.conf           // 配置文件在被控主机的路径
    - user: root                            // nginx 用户名
    - group: root                           // 用户所在组
    - mode: 644                             // 权限
    - template: jinja                       // 配置文件使用 jinja 模板

  service.running:                          //3. 运行服务管理
    - enable: True                          // 可以运行
    - reload: True                          // 可以重载
    - watch:
      - file: /etc/nginx/nginx.conf
      - pkg: nginx
```

其中，salt://nginx/nginx.conf 为配置模板文件位置；enable:True 检查服务是否在开机自启动服务队列中，如果不在则添加上，等价于 chkconfig nginx on 命令；reload:True 表示服务支持 reload 操作，不添加则会默认执行 restart 操作；watch 既检测 /etc/nginx/nginx.conf 是否发生变化，又确保 nginx 已安装成功。

实际上，上面的配置文件和在一台服务器上直接安装 nginx 的过程是相同的，首先是安装包，然后是对配置文件进行修改，最后是启动服务。

（4）使用 jinja 模板定义 nginx 配置文件 nginx.conf，首先创建一个 nginx 目录，因为上面定义了 nginx 配置文件的源路径。

```
[root@master ~]# mkdir /srv/salt/nginx
```

nginx.conf 配置文件可以根据自己的需求进行编写。

```
[root@master ~]# vi /srv/salt/nginx/nginx.conf
user nginx;
worker_processes {{grains['num_cpus']}};
{% if grains['num_cpus'] ==1 %}
worker_cpu_affinity 10;
{% elif grains['num_cpus'] ==2 %}
worker_cpu_affinity 01 10;
{% elif grains['num_cpus'] == 4 %}
worker_cpu_affinity 0001 0010 0100 1000;
{% elif grains['num_cpus'] == 8 %}
worker_cpu_affinity 00000001 00000010 00000100 00001000 00010000 00100000 01000000 10000000;
```

```
{% else %}
worker_cpu_affinity 0001 0010 0100 1000;
{% endif %}
worker_rlimit_nofile {{ grains['max_open_file'] }};
error_log  /var/log/nginx_error.log;
pid        /var/run/nginx.pid;
events
    {
        worker_connections {{ grains['max_open_file'] }};
    }
http
    {
        include     /etc/nginx/mime.types;
        default_type  application/octet-stream;
        sendfile on;
        keepalive_timeout 60;
        log_format  main  '$remote_addr - $remote_user [$time_local] "$request" '
                    '$status $body_bytes_sent "$http_referer" '
                    '"$http_user_agent" "$http_x_forwarded_for"' ;

        server{
          listen 80 default_server;
          server_name _;

          location / {
             root {{ pillar['nginx']['root'] }};
             index index.html index.htm;
          }
          error_page 404   /404.html;
          location = /404.html {
             root /usr/share/nginx/html;
          }

          error_page 500 502 503 504 /50x.html;
          location = /50x.html {
             root /usr/share/nginx/html;
          }
        }
    }
```

（5）现在我们就可以在主控端执行刷新 state 配置命令，让两台被控端去执行安装 nginx 并配置。

```
[root@master ~]# salt '*' state.highstate
web01.saltstack.com:
----------
      ID: nginx                    // 1 安装
```

```
          Function: pkg.installed
           Result: True
           Comment: Package nginx is already installed.
           Started: 01:26:17.824056
          Duration: 2044.66 ms
           Changes:
          ----------
              ID: nginx                        // 2 配置
          Function: file.managed
            Name: /etc/nginx/nginx.conf
           Result: True
           Comment: File /etc/nginx/nginx.conf is in the correct state
           Started: 01:26:19.871986
          Duration: 22.376 ms
           Changes:
          ----------
              ID: nginx                        // 3 启动
          Function: service.running
           Result: True
           Comment: Service nginx is already enabled, and is running
           Started: 01:26:19.895443
          Duration: 180.684 ms
           Changes:
                ----------
                nginx:
                   True

          Summary
          ------------
          Succeeded: 3 (changed=1)
          Failed:    0
          ------------
          Total states run:    3
```

通过执行结果我们看到了三个 ID，它们相当于三个任务，第一个安装，第二个配置，第三个启动。而且显示三个都成功了，失败为零个。如果不放心也可以去被控端看看 nginx 是否启动。

```
[root@web01 ~]# ps -ef | grep nginx
root      9443     1  0 01:26 ?        00:00:00 nginx: master process /usr/sbin/nginx -c /etc/nginx/nginx.conf
nginx     9445  9443  0 01:26 ?        00:00:00 nginx: worker process
root      9500  3934  0 01:29 pts/1    00:00:00 grep nginx
```

我们再看看 web01.saltstack.com 节点的 nginx 主配置文件 nginx.conf。

```
[root@localhost ~]# more /etc/nginx/nginx.conf
user nginx;
worker_processes 1;                    // 根据 cpu 个数配置
worker_cpu_affinity 10;
worker_rlimit_nofile 1024;             // 根据定义的 max_open_file 定义

error_log  /var/log/nginx_error.log;
pid        /var/run/nginx.pid;

events
   {
        worker_connections 1024;       // 根据定义的 max_open_file 定义
   }
// 省略内容
        location / {
            root /data;                // 定义的 nginx 目录
            index index.html index.htm;
        }
// 省略内容
```

查看 web02.saltstack.com 节点的 nginx 主配置文件 nginx.conf，上面注释的内容有相应的变化。

这样基本就完成了通过 SaltStack 批量部署 nginx 并配置。

本章总结

- SaltStack 由 Master 和 Minion 构成，Master 和 Minion 之间是通过 ZeroMQ（消息队列）进行通信的。
- SaltStack 的常用模块有 pkg、file、cmd、user、service、cron。
- 定义的属性值如果是经常变化的，采用 pillar；如果是很固定、不容易改变的，采用 grains。
- state 是 SaltStack 最核心的功能，通过预先定制好的 sls 文件（salt state file），对被控主机进行管理，如程序包、文件、网络配置、系统服务、系统用户等。

随手笔记

第 9 章

自动化运维之 Puppet

技能目标

- 熟悉 Puppet 工作原理
- 能够进行 Puppet 部署与应用

本章导读

随着各种业务对 IT 的依赖渐重,企业的 IT 基础架构规模不断扩张。作为一名系统工程师我们将如何应对这种日益增长的 IT 架构呢?以前系统工程师就像流水线上的一名工人,不断重复地做着同样的工作,现在这一切即将改变,我们将引入运维自动化工具 Puppet。本章首先介绍 Puppet 的工作原理,然后介绍 Puppet 的部署与应用。

知识服务

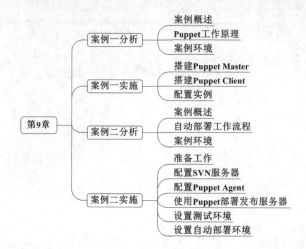

9.1　案例一分析

1. 案例概述

作为一名系统管理员，维护服务器的正常运行是最基本的职责。在管理几台到几十台服务器时，大部分管理员喜欢用自己编写的小工具来维护，但是随着服务器数量的增多，任务量也逐渐增加，这时就需要简洁而又强大的框架来完成系统管理任务。

为了实现这一目的，我们将引入一批工具，这批工具是"可编程"的，系统管理员只需要为这批工具写上几行"代码"，它便会自动完成所有的工作，这批工具就是运维自动化 puppet。在一些大型互联网企业中，运维自动化 puppet 管理着几百甚至上千台服务器，它可以针对多台服务器进行统一操作，如部署统一软件、进行统一上线维护等，而且能够快速完成上线部署，减少人力及误操作风险。

2. 案例前置知识点

Puppet 工作原理如下：

Puppet 的目的是让管理员只集中于要管理的目标，而忽略实现的细节。Puppet 既可以在单机上使用，也可以以 C/S 结构使用。在大规模使用 Puppet 的情况下，通常使用 C/S 结构，在这种结构中 Puppet 客户端只运行 puppeclient，Puppet 服务端只运行 puppetmaster。

具体的工作流程如图 9.1 所示。

（1）客户端 Puppet 调用 facter（facter 是通过 SSL 加密收集及检测分析客户端配置信息的一个工具），facter 探测出主机的一些变量，如主机名、内存大小、IP 地址等。Puppet 把这些信息通过 SSL 连接发送到服务端。

（2）服务端的 puppetmaster 通过 facter 工具分析检测客户端的主机名，然后找到项目主配置文件 manifest 里面对应的 node 配置，并对该部分内容进行解析。facter 发送过来的信息可以作为变量处理，node 牵涉到的代码才进行解析，其他没牵涉的代码不解析。解析分为几个阶段，首先进行语法检查，如果语法没错，就继续解析，解析的结果生成一个中间的"伪代码"，然后把"伪代码"发送给客户端。

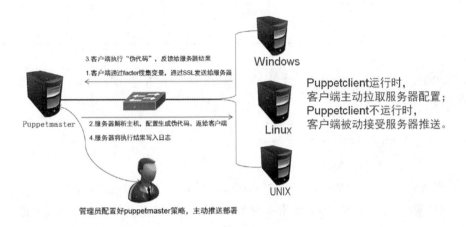

图 9.1　Puppet 工作流程

（3）客户端接收到"伪代码"并且执行，然后把执行结果发送给服务器。

（4）服务端把客户端的执行结果写入日志。

Puppet 工作过程中有以下两点值得注意：

（1）为了保证安全，Client 和 Master 之间是基于 SSL 和证书的，只有经 Master 证书认证的 Client 才可以与 Master 通信。

（2）Puppet 会让系统保持在人们所期望的某种状态并一直维持下去，如检测某个文件并保证其一直存在，保证 ssh 服务始终开启，如果文件被删除了或者 ssh 服务被关闭了，Puppet 下次执行时（默认 30 分钟），会重新创建该文件或者启动 ssh 服务。

3．案例环境

本案例使用四台服务器模拟搭建 Puppet 环境，具体的拓扑如图 9.2 所示。

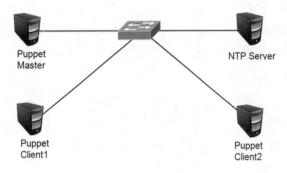

图 9.2　Puppet 实验拓扑

案例环境如表 9-1 所示。

表 9-1 案例环境

主机	操作系统	IP 地址	主要软件
puppetmaster	CentOS 7.3 x86_64	192.168.1.131	puppet-server-3.8.7-1.el7.noarch
puppetclient1	CentOS 7.3 x86_64	192.168.1.132	puppet-3.8.7-1.el7.noarch
puppetclient2	CentOS 7.3 x86_64	192.168.1.137	puppet-3.8.7-1.el7.noarch
NTP Server	CentOS 7.3 x86_64	192.168.1.200	ntp-4.2.6p5-25.el7.centos.1.x86_64

9.2 案例一实施

1. 搭建 Puppet Master

（1）规划服务器主机名

在小规模 Puppet 环境下，一般是修改 /etc/hosts 文件。然而面对上千台服务器，我们需要搭建自己的 DNS 服务器来实现服务通过主机名进行通信，此实验中我们通过修改 /etc/hosts 文件来实现。

```
[root@localhost ~]# hostnamectl set-hostname master.test.cn

[root@localhost ~]# bash
[root@master ~]#

[root@master ~]# vi /etc/hosts
```

添加以下几行：

```
192.168.1.131 master.test.cn
192.168.1.132 client.test.cn
192.168.1.137 client137.test.cn
```

（2）服务器时间同步

由于 Puppet 需要使用 SSL 证书，依赖时间同步，所以需要搭建 NTP 服务器。

1）搭建 NTP Server。

```
[root@localhost ~]# yum -y install ntp
[root@localhost ~]# vim /etc/ntp.conf
```

添加以下两行：

```
server 127.127.1.0
fudge 127.127.1.0 stratum 8
```

其作用是当 /etc/ntp.conf 中定义的 server 都不可用时，将使用 local 时间作为 NTP 服务提供给 NTP 客户端。

```
[root@localhost ~]# systemctl restart ntpd
[root@localhost ~]# systemctl enable ntpd
```

```
[root@localhost ~]# systemctl stop firewalld
[root@localhost ~]# systemctl disable firewalld
[root@localhost ~]# ntpstat
synchronised to local net at stratum 9
   time correct to within 7948 ms
   polling server every 64 s
```

2)puppetmaster 作为 NTP 客户端的配置。

```
[root@master ~]# yum -y install ntpdate
[root@master ~]# ntpdate 192.168.1.200
```

(3)配置 YUM 源

```
[root@master ~]# rpm -Uvh puppetlabs-release-7-12.noarch.rpm
```

(4)安装 Puppet 服务端

```
[root@master ~]# yum install puppet-server
```

(5)启动 Puppet 主程序

生产环境中 firewalld 默认是全部关闭的。

```
[root@master ~]# systemctl stop firewalld
[root@master ~]# systemctl disable firewalld
[root@master ~]# systemctl start puppetmaster.service
[root@master ~]# systemctl enable puppetmaster.service
```

2. 搭建 puppetclient

首先配置 puppetclient1,步骤如下:

(1)规划服务器主机名

```
[root@localhost ~]# hostnamectl set-hostname client.test.cn
[root@localhost ~]# bash
[root@client ~]#
[root@localhost ~]# vi /etc/hosts
```

添加以下几行:

```
192.168.1.131 master.test.cn
192.168.1.132 client.test.cn
192.168.1.137 client137.test.cn
```

确保可以通过域名 ping 通 puppetmaster,即 ping master.test.cn。

(2)服务器时间同步

```
[root@client ~]# yum -y install ntpdate
[root@client ~]# ntpdate 192.168.1.200
 21 Feb 16:07:40 ntpdate[8884]: adjust time server 192.168.1.200 offset -0.000024 sec
```

（3）配置 YUM 源

[root@client ~]#rpm -Uvh puppetlabs-release-7-12.noarch.rpm

（4）安装 Puppet 客户端

[root@client ~]# yum install puppet

（5）修改配置文件

192.168.117.132 和 192.168.117.137 一样，操作如下：

1）修改 client 配置文件。

2）在 [main] 标题下添加以下一行：

[main]
server = master.test.cn   // Puppet Master 的地址
// 以下省略

3）puppetclient2 的配置过程与 puppetclient1 类似，注意将主机名修改为 client137。

（6）申请与注册

Client 端：

分别在 puppetclient1 和 puppetclient2 上进行注册。

[root@client ~]# puppet agent --server=master.test.cn --no-daemonize --verbose

执行完后会有如下提示：

Info: Caching certificate for ca
Info: csr_attributes file loading from /etc/puppet/csr_attributes.yaml
Info: Creating a new SSL certificate request for client.test.cn
Info: Certificate Request fingerprint (SHA256): 9F:EE:A4:EF:89:E7:D4:C2:A6:A0:44:D2:70:5F:86:
    4E:D8:01:97:41:98:FA:AF:0C:32:17:EA:61:E6:3C:6A:ED
Info: Caching certificate for ca

等待一会儿就可以按 Ctrl+C 组合键结束，从 server 查看到申请信息。

Master 端：

查看申请注册的客户端：

[root@master ~]# puppet cert --list
  "client.test.cn"   (SHA256) 9F:EE:A4:EF:89:E7:D4:C2:A6:A0:44:D2:70:5F:86:4E:D8:01:97:41:
    98:FA:AF:0C:32:17:EA:61:E6:3C:6A:ED
  "client137.test.com" (SHA256) 21:7B:02:00:ED:5B:B2:0A:6D:3B:F3:67:F3:0E:02:8E:BA:73:42:2D:
    94:59:31:10:11:33:27:6F:44:2F:1C:50

对未注册的客户端进行注册：

[root@master ~]# puppet cert sign --all
[root@master ~]# puppet cert sign --all
Signing Certificate Request for:
  "client.test.cn" (SHA256) 9F:EE:A4:EF:89:E7:D4:C2:A6:A0:44:D2:70:5F:86:4E:D8:01:97:41:98:
    FA:AF:0C:32:17:EA:61:E6:3C:6A:ED

```
Notice: Signed certificate request for client.test.cn
Notice: Removing file Puppet::SSL::CertificateRequest client.test.cn at '/etc/puppetlabs/puppet/ssl/
    ca/requests/client.test.cn.pem'
Signing Certificate Request for:
  "client137.test.com" (SHA256) 21:7B:02:00:ED:5B:B2:0A:6D:3B:F3:67:F3:0E:02:8E:BA:73:42:2D:
    94:59:31:10:11:33:27:6F:44:2F:1C:50
Notice: Signed certificate request for client137.test.com
Notice: Removing file Puppet::SSL::CertificateRequest client137.test.com at '/etc/puppetlabs/puppet/
    ssl/ca/requests/client137.test.com.pem'
```

可以通过目录去查看已经注册的客户端：

```
[root@master ~]# ll /etc/puppetlabs/puppet/ssl/ca/signed/
total 12
-rw-r--r--. 1 puppet puppet 1960 Feb 22 14:53 client137.test.com.pem
-rw-r--r--. 1 puppet puppet 1952 Feb 22 14:53 client.test.cn.pem
-rw-r--r--. 1 root   root   1988 Feb 21 21:46 master.test.cn.pem
```

此时客户端已经完成证书的请求与签名。

3．配置实例

（1）配置一个测试节点

- 节点信息：/etc/puppet/manifests/nodes/
- 模块信息：/etc/puppet/modules/

案例描述：为了保护 Linux 的 ssh 端口，批量修改客户端 sshd 端口，将端口 22 修改为 9922，并实现重启工作。

创建 ssh 模块，模块下面有三个文件：manifests、templates 和 files。

manifests 里面必须要包含一个 init.pp 的文件，这是该模块的初始（入口）文件，导入一个模块的时候，会从 init.pp 开始执行。可以把所有的代码都写到 init.pp 里面，也可以分成多个 pp 文件，init 再去包含其他文件。定义 class 类名的时候必须是 ssh，这样才能实现调用。

files 目录是该模块的文件发布目录，Puppet 提供一个文件分发机制，类似 rsync 的模块。

templates 目录包含 erb 模型文件，这个和 file 资源的 template 属性有关（很少用）。

Master 端配置：

1）创建必要的目录。

```
[root@release ~]# cd /etc/puppet/
[root@master puppet]# mkdir -p modules/ssh/{manifests,templates,files}
[root@master puppet]# mkdir manifests/nodes
[root@master puppet]# mkdir modules/ssh/files/ssh
[root@master puppet]# chown -R puppet modules/          // 修改权限
```

此时 /etc/puppet/modules/ssh/ 目录下结构为：

```
[root@master puppet]# ll modules/ssh/
total 0
drwxr-xr-x. 3 puppet root 17 Feb 22 15:38 files
drwxr-xr-x. 2 puppet root  6 Feb 22 15:36 manifests
drwxr-xr-x. 2 puppet root  6 Feb 22 15:36 templates
```

2）创建模块配置文件 install.pp。

[root@master ~]# vi /etc/puppet/modules/ssh/manifests/install.pp

输入以下信息（首先确定客户端已安装 ssh 服务）：

```
class ssh::install{
package{ "openssh":
    ensure => present,
   }
 }
```

3）创建模块配置文件 config.pp。

[root@master ~]# vi /etc/puppet/modules/ssh/manifests/config.pp

输入以下信息配置需要同步的文件。

```
class ssh::config{
    file { "/etc/ssh/sshd_config":           // 配置客户端需要同步的文件
            ensure => present,                // 确定客户端中此文件存在
            owner =>"root",                   // 文件所属用户
            group =>"root",                   // 文件所属组
            mode =>"0600",                    // 文件属性
            source =>"puppet://$puppetserver/modules/ssh/ssh/sshd_config",
                                              // 从服务端同步文件
            require => Class["ssh::install"], // 调用 install.pp 确定 ssh 已经安装
            notify => Class["ssh::service"],  // 如果 config.pp 发生变化，则通知 service.pp
        }
}
```

4）创建模块配置文件 service.pp。

[root@master ~]# vi /etc/puppet/modules/ssh/manifests/service.pp

输入以下信息：

```
class ssh::service {
    service {"sshd":
        ensure=>running,            // 确定 ssh 运行
        hasstatus=>true,            // puppet 该服务支持 status 命令，即类似 service sshd status
        hasrestart=>true,           // puppet 该服务支持 restart 命令，即类似 service sshd restart
        enable=>true,               // 服务器是否开机启动
        require=>Class["ssh::config"]  // 确认 config.pp 调用
    }
}
```

5）创建模块主配置文件 init.pp。

[root@master ~]# vi /etc/puppet/modules/ssh/manifests/init.pp

将以上配置文件加载进去，输入以下信息：

```
class ssh{
    include ssh::install,ssh::config,ssh::service
}
```

此时 /etc/puppet/modules/ssh/manifests 目录下有如下四个文件：

```
[root@master ~]# ll /etc/puppet/modules/ssh/manifests
total 16
-rw-r--r-- 1 root root 674 Feb 25 15:08 config.pp
-rw-r--r-- 1 root root  68 Feb 25 15:10 init.pp
-rw-r--r-- 1 root root  85 Feb 25 15:07 install.pp
-rw-r--r-- 1 root root 517 Feb 25 15:09 service.pp
```

6）建立服务端 ssh 统一维护文件。

由于服务端和客户端的 sshd_config 文件默认一样，此时将服务端 /etc/ssh/sshd_config 复制到模块默认路径。

[root@master ~]# cp /etc/ssh/sshd_config /etc/puppet/modules/ssh/files/ssh/
[root@master ~]# chown –R puppet /etc/puppet/modules/ssh/files/ssh/   // 修改权限

7）创建测试节点配置文件，并将 ssh 加载进去。

[root@master ~]# vi /etc/puppet/manifests/nodes/ssh.pp

输入以下信息：

```
node 'client.test.cn'{
    include ssh
}
node 'client137.test.cn'{
    include ssh
}
```

8）将测试节点载入 Puppet，即修改 site.pp。

[root@master ~]# vi /etc/puppet/manifests/site.pp

输入以下信息：

import "nodes/ssh.pp"

9）修改服务端维护的 sshd_config 配置文件。

```
[root@master ~]# vi /etc/puppet/modules/ssh/files/ssh/sshd_config
    Port 9922                    // 修改
    #AddressFamily any
```

```
#ListenAddress 0.0.0.0
#ListenAddress ::
```

10)重新启动 puppet。

```
[root@master ~]# systemctl restart puppetmaster.service
```

（2）客户端主动拉取

一般在小规模自动化集群中，如代码上线需重启服务时，为了防止出现网站暂时性无法访问的问题，每台客户端运行一次 puppet agent -t 命令，选择模式根据客户端集群规模的大小。根据经验，一般 Puppet 服务器到各客户端会建立 ssh 信任，然后自定义 shell 脚本，ssh 批量让客户端执行 Puppet 同步命令。

Client 端：

192.168.1.132 端执行命令如下：

```
[root@client ~]# puppet agent -t
// 省略
+++ /tmp/puppet-file20170225-14206-7n2udv    2017-02-25 15:46:32.537630089 +0800
@@ -14,7 +14,7 @@
# SELinux about this change.
# semanage port -a -t ssh_port_t -p tcp #PORTNUMBER
#
-#Port 22
+Port 9922
#AddressFamily any
#ListenAddress 0.0.0.0
#ListenAddress ::

Info: Computing checksum on file /etc/ssh/sshd_config
// 省略
```

此时命令在 Client 端已经执行成功，验证如下：

```
[root@client ~]#cat /etc/ssh/sshd_config |grep Port
Port 9922
#GatewayPorts no
```

查看服务器 ssh 服务是否重启，端口是否生效。

```
[root@client ~]netstat -tunlp| grep ssh
tcp    0    0 0.0.0.0:9922    0.0.0.0:*    LISTEN    2103/sshd
tcp    0    0 :::9922         :::*         LISTEN    2103/sshd
```

（3）服务器推送同步

当大规模部署时，采用服务器推送模式。

Client 端：

192.168.1.137 端修改：

1）修改配置文件如下：

```
[root@client137 ~]#vi /etc/puppet/puppet.conf
```

最后一行添加如下:

```
listen = true          // 使 puppet 监听 8139 端口
```

2)验证配置文件 auth.conf 定义了一些验证信息及访问权限。

```
[root@client137 ~]#vi /etc/puppet/auth.conf
```

最后一行添加如下:

```
allow *                // 允许任何服务端推送
```

3)启动 Puppet 客户端。

```
[root@client137 ~]# systemctl start puppetagent.service
```

查看 /etc/ssh/sshd_config 的内容如下:

```
#Port 22
#AddressFamily any
#ListenAddress 0.0.0.0
#ListenAddress ::
```

确认启动 ssh 服务。

```
[root@client137 ~]# netstat -tunlp| grep ssh
tcp    0    0 0.0.0.0:22     0.0.0.0:*    LISTEN    3872/sshd
tcp    0    0 :::22          :::*         LISTEN    3872/sshd
```

Master 端:

4)开始往客户端推送。

```
[root@master ~]# puppet kick client137.test.cn
Triggering client137.test.cn
Getting status
status is success
client137.test.cn finished with exit code 0
Finished
```

5)校验结果如下:

此时在 Client 端已经执行成功,验证如下:

```
[root@client137 ~]# cat /etc/ssh/sshd_config |grep Port
Port 9922
#GatewayPorts no
```

查看服务器 ssh 服务是否重启,端口是否生效。

```
[root@client137 ~]# netstat -tunlp | grep sshd
tcp    0    0 0.0.0.0:9922   0.0.0.0:*    LISTEN    2315/sshd
tcp    0    0 :::9922        :::*         LISTEN    2315/sshd
```

9.3 案例二分析

1. 案例概述

如果有工作经验的朋友看到"部署"二字，第一时间肯定会想到的是将新的应用替换或者覆盖旧的应用，然后重启服务之类的。没错，这里的自动部署就是将新的应用自动覆盖旧的应用，起到一个更新的作用，主要还是利用了文件拷贝的原理。

网上大部分自动部署的案例是将 svn 和 puppetmaster 整合在一台服务器上，如果 svn 服务器里面的代码有更新，puppetagent 就会根据自身设置拉取更新文件的时间完成全自动部署。这样做的唯一缺点就是如果出现问题不是能很好的控制，如果由于开发人员误操作导致一个错误的文件已提交到 svn 服务器上，那么 agent 节点也会更新成错误的文件，那么 web 站点将会出现故障，所以这里是将 svn 服务器、puppetmaster 服务器、发布服务器全部单独分开。我们只要保证发布服务器上的代码不会更新错误的文件，web 站点也就不会自动更新。也正是因为这样，我们就不能用 puppetmaster 作为文件服务器，而是以发布服务器作为文件服务器，所以需要用到 Puppet 的 rsync 模块。

2. 自动部署工作流程

自动部署工作流程如图 9.3 所示。

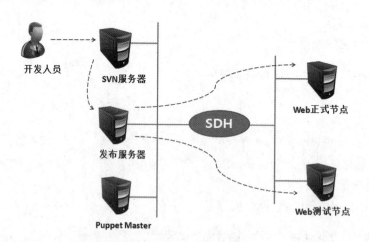

图 9.3　自动部署工作流程

（1）开发人员手工提交更新代码到 svn 服务器上。

（2）开发人员使用系统账号登录到发布服务器上，手工执行与发行版相对应的项目脚本，也就是从 svn 服务器上检测出要更新的代码到发布服务器上。

（3）待脚本执行完后，web 测试环境的应用会根据自身设置的时间自动更新或者更新后执行一些其他动作。这里暂定五分钟。

（4）开发人员和测试人员会进行相关测试。

（5）如果测试环境通过测试，需要发布正式环境的话，请开发人员邮件或者电话通知运维人员要发版的项目。

（6）同样运维人员也是手工执行与发行版相对应的项目脚本（正式环境发布脚本），发布完成后五分钟左右通知开发人员发布结束。

3. 案例环境

本案例使用五台服务器模拟搭建自动部署环境，具体环境如表 9-2 所示。

表 9-2　案例环境

IP 地址	主机名	操作系统版本	用途
192.168.1.131	svn	CentOS7.3（64 位）	svn 服务器
192.168.1.132	release.puppet.com	CentOS7.3（64 位）	发布服务器
192.168.1.133	master.puppet.com	CentOS7.3（64 位）	Puppet Master
192.168.1.138	web.puppet.com	CentOS7.3（64 位）	web 正式节点
192.168.1.139	web-test.puppet.com	CentOS7.3（64 位）	web 测试节点

上述五台服务器除去 svn 不需要上网环境，其余都需要能连接互联网。确保发布服务器与另外四台服务器都是相通的。另外确保所有服务器的时间都同步，这里的自动部署网络环境比较适合于公司内部机房有一条 SDH 专线连接到 IDC 机房的情况，而不是通过互联网去部署，因为这样相当于在局域网内执行动作，速度和安全性都是可靠的。svn 服务器、发布服务器、puppetmaster 都在公司内部，web 正式节点和 web 测试节点都在 IDC 机房中，所以说自动部署也在很大程度上依赖于网络环境。下面就重点介绍一下 web 项目从部署测试环境到部署正式环境是如何操作的。

9.4　案例二实施

1. 准备工作

（1）添加域名解析

需要修改 puppetmaster、发布服务器、web 正式节点、web 测试节点的 hosts 文件，因为 Puppet 都是基于域名的方式来管理节点的。看到表 9-2 中的主机名就知道，如果 IDC 机房内部有 DNS 管理更好，因为以后更多的项目都需要自动部署。

/etc/hosts 文件中增加如下内容：

```
192.168.1.132 release.puppet.com
192.168.1.133 master.puppet.com
192.168.1.138 web.puppet.com
192.168.1.139 web-test.puppet.com
```

（2）处理防火墙和 SELinux

```
[root@master ~]# systemctl stop firewalld
[root@master ~]# systemctl disable firewalld
[root@master ~]# setenforce 0
```

其他节点采用相同配置。

（3）时间同步

因为 Puppet Master 节点需要证书的签发，所以在所有节点上执行同步时间命令，与网络上的时钟服务器进行时间同步，以免客户端获取证书时出错。

```
[root@master ~]# yum install ntpdate
[root@master ~]# ntpdate pool.ntp.org
23 Feb 21:05:12 ntpdate[4449]: adjust time server 115.28.122.198 offset 0.016971 sec
```

其他节点采用相同配置。

（4）安装软件包

在 Puppet Master、发布服务器、web 正式节点和 web 测试节点都需要安装 Puppet 源。Puppet Master 需要安装 puppet-server 包，而发布服务器、web 正式节点和 web 测试节点都被看作是 Puppet 的被管理节点，也就是通常所说的 Agent 端。

```
[root@master ~]# rpm -Uvh puppetlabs-release-7-12.noarch.rpm
```

其他节点配置相同。

1）Master 端

Puppet Master 端安装 Puppet 服务端软件包。

```
[root@master ~]# yum install puppet-server
[root@master ~]# systemctl start puppetmaster.service
[root@master ~]# systemctl enable  puppetmaster
[root@master ~]# netstat -antpu |grep 8140
tcp    0    0 0.0.0.0:8140    0.0.0.0:*       LISTEN    1673/ruby
```

2）Agent 端

发布服务器、web 正式节点和 web 测试节点分别安装 Puppet 客户端软件包。

```
[root@release ~]# yum install puppet
```

web 正式节点和 web 测试节点配置相同。

3）SVN 服务器

SVN 服务器使用本地的 YUM 源即可。

```
[root@svn ~]# yum install subversion
[root@svn ~]# svnserve --version
svnserve, version 1.7.14 (r1542130)
   compiled Nov 20 2015, 19:25:09
// 以下信息略
```

2. 配置 svn 服务器

（1）创建 SVN 版本库目录

```
[root@svn ~]# mkdir -p /var/svn/html/
```

（2）创建版本库

```
[root@svn ~]# svnadmin create /var/svn/html/
```

执行此命令后会生成如下文件：

```
[root@svn ~]# ls /var/svn/html/
conf  format  locks
db    hooks   README.txt
```

（3）修改 SVN 配置文件

```
[root@svn ~]# vi /var/svn/html/conf/svnserve.conf
```

修改如下信息：

```
 anon-access = read                              // 匿名用户可读
auth-access = write                              // 授权用户可写
 password-db = /var/svn/html/conf/passwd         // 指定账号文件
 authz-db = /var/svn/html/conf/authz             // 指定权限文件
```

（4）设置账号密码

```
[root@svn ~]# vi /var/svn/html/conf/passwd
[users]
```

添加如下内容：

```
alpha = alpahapasswd
sysadmin = sysadminpasswd
```

（5）设置权限

```
[root@svn ~]# vi /var/svn/html/conf/authz
```

添加如下内容：

```
[/]
sysadmin = r
alpha = r
[web]
sysadmin =r
alpha =rw
```

（6）启用版本库

```
[root@svn ~]# svnserve -d -r /var/svn/html
```

（7）创建测试目录

```
[root@svn ~]# cd /var/svn/html/
[root@svn html]# mkdir web
[root@svn html]# svn import web file:///var/svn/html/web -m "init svn"
```

会有如下提示信息：

```
Committed revision 1.
```

3. 配置 Puppet Agent

（1）修改配置文件

因为 Puppet Master 管理客户端 Agent 是基于 SSL 方式通过证书通信，所以需要在所有的 Agent 角色上，也就是在发布服务器、web 正式节点、web 测试节点上修改 Puppet 的主配置文件 /etc/puppet/puppet.conf，在 [main] 字段中指向服务端。

```
[root@release ~]# vi /etc/puppet/puppet.conf
[main]
```

添加如下内容：

```
server = master.puppet.com             // Puppet Master 的地址
```

启动 Agent 端服务：

```
[root@release ~]# systemctl start puppetagent.service
```

在 web 正式节点和 web 测试节点上做相同操作。

（2）申请证书

然后在所有的 Agent 角色上分别执行如下命令：

```
[root@release ~]# puppet agent --server=master.puppet.com  --verbose --no-daemonize
```
执行完后会有如下提示：
```
Info: Caching certificate for ca
Info: csr_attributes file loading from /etc/puppet/csr_attributes.yaml
Info: Creating a new SSL certificate request for release.puppet.com
Info: Certificate Request fingerprint (SHA256): 1E:92:5D:9C:59:8A:96:19:82:DB:54:79:B8:75:8F:92:
    6E:BE:20:BC:7D:1A:37:A0:46:8E:22:0A:C7:9D:C9:A4
Info: Caching certificate for ca
```

在 web 正式节点和 web 测试节点上做相同的操作，然后在 Puppet Master 查看等待被签名的证书的客户端。

```
[root@master ~]# puppet cert –list
```

执行完后会有如下显示：

```
"release.puppet.com"  (SHA256) 1E:92:5D:9C:59:8A:96:19:82:DB:54:79:B8:75:8F:92:6E:BE:20:BC:
    7D:1A:37:A0:46:8E:22:0A:C7:9D:C9:A4
"web-test.puppet.com" (SHA256) F6:83:51:F4:F4:AD:DB:12:CF:D7:7F:05:C6:74:50:B6:0A:25:02:
    2E:AD:28:5A:E6:BB:B8:58:AD:07:3D:EE:44
```

"web.puppet.com"　　(SHA256) 42:05:95:43:AD:0F:F2:DF:7F:BE:DB:B3:E9:61:0D:85:15:CF:A9: 75:06:FC:BC:E7:98:1E:25:A2:94:20:30:81

再在 Puppet Master 执行如下命令完成签名：

```
[root@master ~]# puppet cert sign release.puppet.com
[root@master ~]# puppet cert sign web.puppet.com
[root@master ~]# puppet cert sign web-test.puppet.com
```

如果 Agent 端需要被签名的证书很多，可以执行 puppet cert --sign -all 完成签名。

4. 使用 Puppet 部署发布服务器

（1）下载安装 concat 和 rsync 模块

下面开始在发布服务器、puppetmaster 上都需要下载安装 concat 和 rsync 模块，如果发布服务器上不下载模块，一会儿发布服务器上生成的 rsyncd.conf 文件里面会是空的，Puppet 的 rsync 模块在 GitHub 位置 https://github.com/onyxpoint/pupmod-rsync。这里是使用 git，如果服务器没有安装，可以先使用 yum -y install git 来安装；如不想安装 git 也可以下载 zip 压缩包，解压到相应目录。

在发布服务器、Puppet Master 上执行如下命令：

```
[root@release ~]# yum install -y git
[root@release ~]# cd /etc/puppet/modules
[root@release modules]#git clone https://github.com/onyxpoint/pupmod-concat && mv
    pupmod-concat concat
[root@release modules]#git clone https://github.com/onyxpoint/pupmod-rsync && mv
    pupmod-rsync rsync
```

上述两个命令都是下载后重新命名为新的模块名。

```
[root@release modules]# ll
total 0
drwxr-xr-x 5 root root  72 Feb 25 10:30 concat
drwxr-xr-x 8 root root 118 Feb 25 10:30 rsync
```

Puppet Master 上的操作相同。

（2）配置 Puppet Master

Puppet Master 开始创建管理 Agent 节点目录及文件，方便以后管理。

```
[root@master ~]# mkdir -p /etc/puppet/manifests/nodes
[root@master ~]# vi /etc/puppet/manifests/nodes/release.puppet.com.pp
```

添加内容如下：

```
class rsync::client inherits rsync {
}

node 'release.puppet.com' {
include 'rsync::server'
rsync::server::global { 'global':
  address => '192.168.1.132'
```

```
    }

    rsync::server::section { 'web':
      comment => 'This is formal file path',
      path => '/var/www/html/web',
      hosts_allow => '192.168.1.138'
    }

    rsync::server::section { 'web_test':
      comment => 'This is test file path',
      path => '/var/www/html/web_test',
      hosts_allow => '192.168.1.139'
    }
    ......... 此处以后可以再添加别的项目的配置。
    }
```

然后再创建一个 site.pp 配置文件，定义 Puppet 相关的变量和默认配置，是 Puppet 最先读取的文件。

[root@master ~]# vi /etc/puppet/manifests/site.pp

添加如下内容：

import 'nodes/release.puppet.com.pp'

（3）自动配置发布服务器

在发布服务器上执行如下命令，最后加上 --debug 参数用于调试输出信息。

[root@release ~]# puppet agent --server=master.puppet.com --test -v

以下是部分信息输出，正常执行后返回结果如下：

Notice: /Stage[main]/Rsync::Server/Concat_build[rsync]/target: ["global", "*.section"] used for ordering
Notice: /Stage[main]/Rsync::Server/File[/etc/rsyncd.conf]/mode: mode changed '0644' to '0400'
Notice: /Stage[main]/Rsync::Server/File[/etc/rsyncd.conf]/content: audit change: newly-recorded value {md5}2d46f901df5bf5457994f556eca1aad1
Info: /Stage[main]/Rsync::Server/File[/etc/rsyncd.conf]: Scheduling refresh of Service[rsync]
Notice: /Stage[main]/Rsync::Server/Service[rsync]/ensure: ensure changed 'stopped' to 'running'
Info: /Stage[main]/Rsync::Server/Service[rsync]: Unscheduling refresh on Service[rsync]
Info: Creating state file /var/lib/puppet/state/state.yaml
Notice: Finished catalog run in 4.32 seconds

查看后发现发布服务器的 /etc 目录下已经自动生成 rsyncd.conf 文件，但是 rsyncd 服务是未启动的。

自动生成的 /etc/rsyncd.conf 文件内容如下：

[root@release ~]# cat /etc/rsyncd.conf
pid file = /var/run/rsyncd.pid
syslog facility = daemon

```
port = 873
address = 192.168.1.132
[web]
comment = This is formal file path
path = /var/www/html/web
use chroot = false
max connections = 0
max verbosity = 1
lock file = /var/run/rsyncd.lock
read only = true
write only = false
list = false
uid = root
gid = root
outgoing chmod = o-w
ignore nonreadable = true
transfer logging = true
log format = "%o %h [%a] %m (%u) %f %l"
dont compress = *.gz *.tgz *.zip *.z *.rpm *.deb *.iso *.bz2 *.tbz *.rar *.jar *.pdf *.sar *.war
hosts allow = 192.168.1.138
hosts deny = *
[web_test]
comment = This is test file path
path = /var/www/html/web_test
use chroot = false
max connections = 0
max verbosity = 1
lock file = /var/run/rsyncd.lock
read only = true
write only = false
list = false
uid = root
gid = root
outgoing chmod = o-w
ignore nonreadable = true
transfer logging = true
log format = "%o %h [%a] %m (%u) %f %l"
dont compress = *.gz *.tgz *.zip *.z *.rpm *.deb *.iso *.bz2 *.tbz *.rar *.jar *.pdf *.sar *.war
hosts allow = 192.168.1.139
hosts deny = *
```

接下来手动启动 rsync 服务。

```
[root@release ~]# rsync --daemon
```

查看 873 端口是否开启，如果发布服务器上开启防火墙记得打开 873 端口。

```
[root@release ~]# netstat -antpu |grep rsync
tcp        0      0 192.168.1.132:873        0.0.0.0:*        LISTEN      2219/rsync
```

如果想下次发布服务器能自动启动 rsync 服务，那么需要修改 Puppet Master 的 rsync 模块文件，将 /etc/puppet/modules/rsync/manifests/server.pp 文件中的如下行注释掉：

stop => "/bin/kill 'cat \\'grep \"pid file\" /etc/rsyncd.conf | cut -f4 -d' \\'"",

新增如下所示：

stop => "/bin/kill 'ps -ef | grep rsync | grep -v grep | awk '{print $2}'"",

5. 设置测试环境

（1）编写测试脚本

接下来在发布服务器上编写 web 项目测试环境发布脚本，内容大概如下：

```bash
[root@release ~]# vi webtest.sh
#web 测试环境版本发布工具
#!/bin/bash
# 定义路径
path=$(cd $(dirname "$0");pwd);
svnRoot="svn://192.168.1.131/webtest";
svndir="/var/svn/html/web_test";
webdir="/var/www/html/web_test";

choose="no yes"
echo " 请确认要发布的是 web 项目测试环境 ";
select comfirm in $choose
do
  if [  "${comfirm}" != "yes"  ];then
        echo " 发布程序结束 "
        exit 0;
done;
echo " 请确认 SVN 地址 ( 默认 no):${svnRoot}";
select comfirm in $choose
do
  if [  "${comfirm}" != "yes"  ];then
        echo " 发布程序结束 "
        exit 0;
    fi
    break;
done;

# 开始检出程序代码
svn co ${svnRoot}   ${svndir};
rsync -acvz --exclude=".svn/" --delete --delete-after ${svndir}/*  ${webdir}
echo " 覆盖配置 "
```

脚本里面还可以增加其他的预设动作，根据不同的业务逻辑进行修改，手工执行脚本检出 SVN 代码到发布服务器上。

注意执行脚本之前需要在发布服务器上创建目录。

```
[root@release ~]# mkdir -p /var/www/html/web_test
[root@release ~]# mkdir -p /var/svn/html/web_test
```

（2）配置 web 测试环境客户端

接下来需要配置 web 测试环境客户端，在 Puppet Master 上定义客户端模板文件。

```
[root@master ~]# mkdir /etc/puppet/modules/rsync/manifests/client
```

在 client 目录下新建一个 host.pp 文件，这里 client 目录是新建的。

```
[root@master ~]# vi /etc/puppet/modules/rsync/manifests/client/host.pp
```

添加如下内容：

```
define rsync::client::host ($title,$rsyserver,$source,$target){
    exec{"$title":
    path => "/usr/bin:",
    command => "rsync -acvz --delete $rsyserver::$source $target"
    }
}
```

上述参数含义如下：

title：定义的是一个主题，可以任意定义。

rsyserver：是 rsync server 的 IP。

source：是 rsync server 上定义的目录。

target：是网站测试节点上放置的目录。

（3）创建站点文件

最后在 Puppet Master 上创建一个 web 项目测试环境站点的文件。

```
[root@master ~]# vi /etc/puppet/manifests/nodes/web-test.puppet.com.pp
```

添加如下内容：

```
node 'web-test.puppet.com' {
        include 'rsync::client'
        rsync::client::host {"web 项目测试环境 ":
        title => 'web 项目测试环境 ',
        source =>'web_test',
        rsyserver => '192.168.1.132',
        target =>'/var/www/html/web_test'
        }
}
```

上述内容很少，因为完全就是调用 rsync 命令复制文件，当然里面也可以增加同步完后的一些预设动作，比如 exec 执行外部脚本之类的。最后需要将 web-test.puppet.com 文件也导入到 site.pp 文件里面。

```
[root@master ~]# echo "import 'nodes/web-test.puppet.com.pp' ">>/etc/puppet/manifests/site.pp
```

（4）手工测试

在 web 项目测试站点上创建同步目录，然后执行如下命令确认发布服务器上的更新文件是否被同步到 web 项目测试站点。

```
[root@web-test ~]# mkdir -p /var/www/html/web_test
[root@web-test ~]# puppet agent --server master.puppet.com --test --debug
```

6. 设置自动部署环境

上述都是手工执行拉取动作，而本案例的目的是自动部署，所以我们需要配置 Agent 自动更新应用。

（1）修改 Agent 自动更新

在 web 项目的测试和正式环境上的 Agent 客户端的 /etc/puppet/puppet.conf 配置文件中的 [agent] 字段中增加如下内容：

```
listen = true
runinterval = 300s          // 表示 5 分钟自动更新
```

然后在 web 项目的测试和正式环境上启动 puppet 服务。

```
# systemctl start puppetagent.service
```

至于发布服务器上为了防止 rsync 消失，可以一小时自动拉取一次。具体时间根据需求而定。

（2）编写正式发布脚本

上述测试环境自动部署成功，需要部署正式环境只需要写一个发布正式环境的脚本，将发布服务器上的测试应用拷贝到发布服务器上的正式应用下面，然后让 web 正式环境节点去自动更新。

```
[root@release ~]# vi web_formal.sh
#web 项目正式环境版本发布工具（此脚本只能测试环境测试通过后运维人员去执行）
#!/bin/bash

# 设置根目录
SOURCE_ROOT='/var/www/html';
SHELL_ROOT=$(cd $(dirname "$0"); pwd);

# 设置源码目录
SOURCE_WEB="${SOURCE_ROOT}/web";
SOURCE_WEB_TEST="${SOURCE_ROOT}/web_test";

choose="no yes"
echo " 请确认要发布的是 web 项目正式环境？（该脚本只能由运维人员执行）";
select comfirm in $choose
do
    if [ "${comfirm}" != "yes" ]; then
```

```
            echo " 发布程序结束 ";
            exit 0;
        fi
    break;
done;

echo " 请确认是否同步正式环境？请务必确认目录的正确性 !";
select COMFIRM in ' 否 ' ' 是 '
do
  if [ "${COMFIRM}" == " 是 " ]; then
    echo ' 同步源码目录 ';
    rsync -avr --delete-after ${SOURCE_WEB_TEST}/* ${SOURCE_WEB};
  fi
  break;
done;
echo 'web 项目正式环境发布完成，请通知开发和测试人员五分钟后进行测试 !'
```

（3）创建节点文件

在 Puppet Master 端创建 web 正式环境的节点文件。

```
[root@master ~]# vi /etc/puppet/manifests/nodes/web.puppet.com.pp
```

文件内容如下：

```
node 'web.puppet.com' {
        include 'rsync::client'
        rsync::client::host {"web 项目正式环境 ":
        title => 'web 项目正式环境 ',
        source =>'web',
        rsyserver => '192.168.1.132',
        target =>'/var/www/html/web'
        }
}
```

同样最后别忘记了将 web.puppet.com.pp 文件 import 到 site.pp 文件里面。

```
[root@master ~]# echo "import 'nodes/web.puppet.com.pp'">>/etc/puppet/manifests/site.pp
```

（4）设置正式节点同步目录

在 web 正式节点上创建相应的同步目录。

```
[root@web ~]# mkdir -p /var/www/html/web
```

本章总结

- Puppet 既可以在单机上使用，又可以以 C/S 结构使用。在大规模使用 Puppet 的情况下，通常使用 C/S 结构，在这种结构中 Puppet 客户端只运行

puppetclient，Puppet 服务端只运行 puppetmaster。
- Puppet 使用 SSL 证书依赖于时间同步，需要搭建 NTP 服务器。
- 通过一个配置实例批量修改客户端 sshd 端口。
- 使用 Puppet 实现代码发布的自动部署。

本章作业

1. 画图并描述 Puppet 的工作流程。
2. NTP 客户端如何与 NTP 服务器做定时同步？
3. 如果启用 firewalld 防火墙，如何编写防火墙规则？